Praise for
THE BRAIN IN CONTEXT

"Jonathan D. Moreno and Jay Schulkin provide a wonderfully engaging account of the human brain and how it works. They smoothly blend the latest in scientific discoveries with a broad historical and philosophical perspective. *The Brain in Context* goes down easy and fires the imagination. Delightful and illuminating!"

Kent Berridge, James Olds Distinguished University Professor of Psychology and Neuroscience, University of Michigan

"Moreno and Schulkin deftly deploy an irresistible array of fascinating stories and witty, accessible analysis to open up a new worldview of our poorly understood brains. They treat us to the very best in philosophical, historical, and scientific analysis and illustrate how we arrived at knowing what we know as well as what we only think that we know about our brain as a whole. *The Brain in Context* is an indispensable read for all of us who use our brains yet scratch our heads about how they work. A true treat for all our minds to read and to ponder about!"

Amy Gutmann, president and Christopher H. Browne Distinguished Professor of Political Science and professor of communication, University of Pennsylvania

"*The Brain in Context* is a grand pragmatic tour through neuroscience and our quest to understand the brain. Remarkable in its trajectory from historical times to the pressing issues of contemporary research, the style is informal and clear. This new view on the state of modern neuroscience will be a pleasure to read for experts and enthusiasts alike, leaving us with a broad sense of the present state of the field."

Michael Hawrylycz, Allen Institute for Brain Science

"Immerse yourself in this exhilarating book and you will be guided skillfully through much of contemporary neuroscience and its associated philosophy; there's always been a close connection between the two. We are our brains: they make us human but also individuals. Many of the most serious and life-damaging disorders occur when the brain malfunctions. In this book, you'll find a most readable account not only of what we currently know about the brain in all its complexity and variety, but also what we don't."

Joe Herbert, University of Cambridge

"*The Brain in Context* is a user-friendly book filled with interesting and informative essays about our brains as we understand them, how we achieved that understanding, and mistakes made along the way. A delight to read."

Joseph LeDoux, author of *The Deep History of Ourselves:*
The Four-Billion-Year Story of How We Got Conscious Brains

"A highly engaging introduction to what we know and don't know about the brain, surveying areas of solid understanding while also emphasizing the large gaps and uncertainties that remain. This book will provide value for anyone seeking a balanced perspective on this fascinating organ, which still remains mysterious in many ways. The authors are distinguished behavioral neuroscientists with broad interests, and they write with verve as well as authority."

Hal Pashler, editor in chief of *Encyclopedia of the Mind*

"How the brain works is as hard to explain to the general public as it is to students studying for a PhD in neuroscience. That's because no one has come up with a 'Grand Theory' of brain function, a conceptual framework analogous to Darwin's theory of biological evolution. *The Brain in Context* serves up an accessible and fascinating introduction to current thinking about what the brain does and how it does it. Moreno and Schulkin have rich backgrounds in writing about neuroscience, and do an admirable job of guiding readers between the current extremes of 'neurohype' (neuroscience is about to solve the world's problems) and 'neuropessimism' (the brain is so complex that tangible neuroscience solutions are indefinitely far away)."

Larry W. Swanson, author of *Brain Architecture:*
Understanding the Basic Plan

THE BRAIN IN
CONTEXT

THE
BRAIN
IN CONTEXT

A Pragmatic Guide to
Neuroscience

JONATHAN D. MORENO
and JAY SCHULKIN

COLUMBIA UNIVERSITY PRESS   *NEW YORK*

Columbia University Press
Publishers Since 1893
New York   Chichester, West Sussex
cup.columbia.edu
Copyright © 2020 Columbia University Press
All rights reserved

Library of Congress Cataloging-in-Publication Data
Names: Moreno, Jonathan D., author.
Title: The brain in context : a pragmatic guide to neuroscience /
Jonathan D. Moreno and Jay Schulkin.
Description: New York : Columbia University Press, [2019] |
Includes bibliographical references and index.
Identifiers: LCCN 2019006025| ISBN 9780231177368 (cloth : alk. paper) |
ISBN 9780231547109 (e-book)
Subjects: LCSH: Neurosciences—Popular works. |
Neurotechnology (Bioengineering)
Classification: LCC RC341 .M727 2019 | DDC 612.8/233—dc23
LC record available at https://lccn.loc.gov/2019006025

Cover design: Alex Camlin
Cover image: Freepik

ALSO BY JONATHAN D. MORENO

Impromptu Man: J. L. Moreno and the Origins of Psychodrama,
Encounter Culture, and the Social Network

Mind Wars: Brain Science and the Military in the Twenty-First Century

The Body Politic: The Battle Over Science in America

Undue Risk: Secret State Experiments on Humans

Deciding Together: Bioethics and Moral Consensus

COAUTHORED BOOKS

Ethics in Clinical Practice (with Judith Ahronheim and Connie Zuckerman)

Discourse in the Social Sciences: Strategies for Translating Models of Mental Illness
(with Barry Glassner)

ALSO BY JAY SCHULKIN

Reflections on the Musical Mind: An Evolutionary Perspective

Naturalism and Pragmatism

Effort: A Behavioral Neuroscience Perspective on the Will

Bodily Sensibility: Intelligent Action

Roots of Social Sensibility and Neural Function

The Neuroendocrine Regulation of Behavior

COAUTHORED BOOKS

Milk: The Biology of Lactation (with Michael L. Power)

The Evolution of Obesity (with Michael L. Power)

CONTENTS

INTRODUCTION

I n sheer weight, our brains aren't all that big—only about three pounds. Sperm whales' brains are four times as large. Does size matter? Not as much as neuroscientists once thought. Some early humans had brains as big as ours, but they weren't as round. Does shape matter? It seems to, but exactly how is a question that remains for neuroscientists to explore. Big or small, round or flat, local or extended, what we can say is that, compared to other common physical objects, our brains punch way above their weight. The apparent disparity between size and functionality has led people to wonder whether we could do more with what we've got.

There's an old joke about a man at a party chatting with a neuroscientist. Expressing his fascination with neuroscience, he says, "You know, I wish you people could figure out how I can use more than ten percent of my brain."

And that's the joke.

OK, it's not funny, and it's not even old. We just made it up. What is true is that the old saw about 10 percent is false. In fact, all of our brain is always at work (though it might not be very efficient). For millennia, people have been trying to figure out which parts of the brain do what. As we will explain, this approach is more promising than just focusing on size, but it's far from the end of the story. More recently, the rush of popular interest in the brain has produced a number of books about the (fill in the blank) brain: *ethical, political, sleeping,* and that perennial favorite, *sexual.* What concerns us is that these books tend to distract the nonscientist reader from the brain as a whole. We believe that the secret to understanding how our brains do as well as they do (and why they don't always do so well) is to see the brain in context.

The *Whole Earth Catalog* (1968) was perhaps as famous and influential for its first cover as for its contents; the issue featured a NASA photo of the planet Earth, seen in its wholeness for the first time. Today there also are many pictures of the "whole brain," including an impressive 3-D digital reconstruction of a human brain by Europe's Human Brain Project and a map of the developing human brain by the Howard Hughes Medical Institute.[1] But even that breakthrough doesn't capture the brain in its true wholeness. For that, you have to take into account the way the brain extends into every part of the body, its genetics, its chemistry, its evolution, its environment from conception to grave, and the way it manifests itself in every aspect of human life.

In 1950, neuroscientist Warren S. McCullough published a paper with the equivocal title "Why the Mind Is in the Head."[2] Even if that is true about the mind—and we don't think it is—it sure isn't true about the brain. In fact, the brain is all over the body, in the form of neural tissues. That's one of the reasons it's hard to study the brain. Unlike so many other solid organs, it can't be isolated in even a rudimentary sense. A brain in a dish isn't really a brain, because the brain isn't simply localized, it's extended. That's yet another reason that mere brain size or shape don't tell us much. These are also the keys to modifying the brain through experience, drugs, or devices. In a nutshell, that's the basic idea of what we have tried to do in this book, to see the brain in its wholeness, in as many ways as we could juggle, always with an eye toward how experience and our interventions change it.

In truth, we understand very little about the whole brain—maybe something like 5 percent of it (a very imprecise calculation, without knowing the denominator!)—or the way it works. So we didn't call this the "whole brain book." That would take a much, much longer book, and not enough is known to justify such a title anyway. But in a single, reasonably sized book about the brain, it is possible to explain the way theories of the brain have appeared, disappeared, and left traces that have come up again.

As a philosopher who has immersed himself in neuroscience and a neuroscientist with intellectual roots in philosophy, we can't help but be interested in the way a theory of the brain is gradually being constructed. We think of theories as guides to experiments and experiments as correctives to theories. The Austrian scientist and philosopher Otto Neurath famously compared human knowledge to a ship being rebuilt at sea: "By using the old beams and driftwood the ship can be shaped entirely anew, but only by gradual reconstruction."[3] In neuroscience, we often seem to be at sea without a vessel, trying to build one with the flotsam and jetsam that rolls by while we paddle furiously, just barely keeping our brain-filled heads above water.

In this book, then, we also assess the future and the past of learning about the brain while we are at sea. Why, over the centuries, did people think the way they did about the brain? From the ancient world to our own time, what experiences have most impressed physicians and led them to think about the brain in certain ways? How did the often unexpected results of medical procedures over thousands of years lead to modern neuroscience?

How have scientific theories about the nature of the brain succeeded and how have they failed? What are modern applications of neurotechnologies teaching us about the brain, both to heal and (as some hope) to improve it, and what challenges remain? As Neurath taught, theories don't need to be born complete; indeed, we should be suspicious of those that are. Rather, theory construction in any field of science is a result of experience with applying what we think we know.

Neuroscience is making great strides, both at the level of theory and in the laboratory. But, as has so often been true in brain science, the experiments are driving the theory. For those waiting for cognitive enhancements, one of the messages of this book is that we shouldn't expect as much from brain interventions as we are often led to believe—certainly not in the short run, and perhaps not ever. In writing this book, a guiding principle has been that we should step gingerly between exaggerating the potency of neurotechnologies ("neurohype") and minimizing it ("neuropessimism"). Tens of millions of years of messing around with one combination of molecules or another is hard to beat. Our reading of that natural history leads us to conclude that there will be as many disappointments as successes, though successes there will be. And there will be disconcerting results as well, both for science and society. New findings will cause us to move in unpredictable directions.

We doubt, for example, that the old argument between those who believe that the mind is reducible to the brain and those who uphold the qualitative independence of mental activity can be resolved by neuroscience. Rather, we suspect that new science will give rise to new philosophical frameworks, as has been the case in the past. At the moment, the most prominent option may be Giulio Tononi's integrated information theory, but critics argue that merely integrated information isn't enough to give rise to consciousness. These arguments shouldn't give rise to despair. The pragmatist Charles Peirce argued that these kinds of disagreements help us to better understand what we're disagreeing about, which is an earmark of science itself. That's one of the reasons that we also call this book a "pragmatic guide to neuroscience." Philosophical problems aren't so much solved as gotten over, or we return to them when we can address them with more depth and understanding. And, ontological debates aside, although spectacular experiments do not in themselves amount to the grand theory of the brain that has been sought for so long, at least in the short run some remarkable things are coming out of neuroscience labs all over the world. Along with a surprising amount of ancient knowledge, taken in context, it all adds up to a glimpse of what the whole brain must be like.

1

ELECTRIFYING

I n Seattle, Washington, scientists transmit signals from one
person's brain to another over the internet to control the second person's hand motions.

In Berkeley, California, researchers collect brain signals
from people watching a video and reconstruct the images in
a computer-generated movie.[1]

In Zurich, Switzerland, neuroscientists give people extra shots
of a brain chemical called oxytocin to make them more trusting
in social situations.[2]

In Tokyo, Japan, tiny electrical stimulation is shown to help
stroke patients recover their ability to swallow.[3]

These twenty-first-century experiments are striking in many
ways, and they often use new technologies to create or to measure
an effect on the brain. But provocative brain experiments aren't
new, and they needn't be high tech. In 1963, armed only with a
radio-equipped transmitter, an audacious Spanish neuroscientist named José Delgado stepped into a bullring in Cordoba to
face an animal bred to fight. As the bull charged at Delgado, he
pushed a button on the device he called a stimoceiver, which
activated an electrode implanted in the bull's brain, causing it to
stop dead in its tracks, but it was otherwise unharmed. Not

satisfied to perform this feat of scientific courage once that day, Delgado conducted the same demonstration with bull after bull. Each one halted in its tracks as the button was pushed.[4]

NO BULL

Delgado, a professor in New Haven and Madrid, was a colorful and controversial pioneer of brain stimulation by electrical impulses. Delivered to regions of the brain like the hypothalamus, these subtle charges could alter, stop, or excite behavior. For years, Delgado worked on placing electrodes in the brain as a means of controlling thought and action. Experimenting with various species, he was particularly interested in controlling violence by stimulating brain regions. Much of that research would never be allowed now, particularly his work with chimpanzees, and the experiments played into ideas about brainwashing and mind control that were so prevalent during the Cold War. A front-page story the *New York Times* called Delgado's bullring experiment "the most spectacular demonstration ever performed of the deliberate modification of animal behavior through external control of the brain."[5] Delgado went so far as to patent an electronic system that he and his colleagues invented for the activation or inhibition of electrical activity in the brain.

Delgado's critics noted that the bull might just have been made dizzy or been incapacitated, but the spectacle made a deep impression on observers as well as on the bull. The drama of Delgado's technical feat overshadowed the question of how it was actually achieved or the precise brain mechanisms that enabled it. Delgado's use of brain stimulation was one of the two most important traditional tools for studying how the brain controls behavior. The other is lesioning—observing the results of damaged brain tissue or creating an injury at what is thought to be a strategic site.

The timing of Delgado's dramatic demonstration was auspicious. The world of post–World War II science was an ambitious one in which everything seemed possible. After all, the secrets of the atom, of the genome, and even of antibiotics had been discovered. Surely the brain, too, would yield up its secrets. But by the 1960s, scientists and clinicians who had hoped for a comprehensive theory of the way the brain works must have felt like they were hitting a wall. Their own work was leading them to results that only made the picture more complicated than it might have seemed originally.

Currently, all over the world, a number of ambitious efforts are under way that might ultimately contribute to "theorizing" the brain, especially the Brain Research Through Advancing Innovative Neurotechnologies (BRAIN) initiative in the United States and the Human Brain Project in Europe. A critical part of these projects will be to gather and analyze vast quantities of data, not unlike the Human Genome Project. One difference between the Human Genome Project and these new brain projects, however, is that the genome project yielded a large amount of data that is now being analyzed. The same cannot be said, as yet, about the brain projects. Moreover, compared to the human genome—as formidable and complex a subject as that was—the brain presents far more targets for the acquisition of new information, which will require a vast array of types of equipment and experiments. And the genome project didn't have to worry about concepts like "mind" and "consciousness," which must somehow be taken into account in brain science, though they present profound philosophical challenges that might never be fully resolved.

But do they need to be resolved in order for a complete theory of the brain to be achieved? Could even a complete brain theory solve them? Experiments like Delgado's, while in some ways gross and uncontrolled, are provocative reminders that the history of neuroscience has largely been light on theory and heavy on demonstration, building a picture of the brain from the ground up rather than from the top down.

SINGING THE BODY ELECTRIC

José Delgado's highly publicized encounter with the bull built on more than a century of fascination with electricity and the nature of life. "I sing the body electric," wrote Walt Whitman in 1867 (in a very different context), but his was not a solo performance. By then, many had electricity on their minds. For Americans, the electric excitement was as old as the republic and closely tied to that history. The statesman-scientist Benjamin Franklin surmised that lightning was a natural form of electricity. When experimenters like Michael Faraday, Georg Simon Ohm, and James Clerk Maxwell analyzed electromagnetism, they helped write a new chapter of the industrial revolution that shaped the young nation. As these scientific developments were unfolding, so were literary renderings in the "romantic science" movement. The visionaries included a very young Mary Shelley, whose *Frankenstein* (1818) featured a stereotypically overambitious scientist (her subtitle was *The Modern Prometheus*), for whom electricity represented the gift of life itself. Though lately we worry about the "two cultures" of science and the humanities, electricity brought those cultures together.

Hundreds of years earlier, Jewish mythology featured the golem, a monster magically generated from clay. The science fiction writer Arthur C. Clarke famously said, "Any sufficiently advanced technology is indistinguishable from magic." The control of electricity to stimulate nonliving tissue made it possible to create something like a golem, life from nonlife. The idea was later called "galvanizing," in homage to Luigi Galvani, who in the eighteenth century electrified the nervous system. Modern-day cardiac pacemakers are a realization of nineteenth-century breakthroughs in technology, such as Alessandro Volta's electric battery of 1799, and some harrowing experiments on dead and living bodies. In those days, the golem and Frankenstein's monster indeed seemed to be coming to life in the laboratory.

Fictional monsters remain a stimulating way to think about what it is to be a human being, and when combined with the science of the day, fiction can help illuminate what is understood about science up to that time and what is not. As we will see here, philosophers use all sorts of fantastic scenarios to reimagine mental activity. But, as we will also explain, the grunt work of brain science itself usually has not relied on studies of the brains of monsters (like Abby Normal's brain in the Mel Brooks film *Young Frankenstein*) but on studies of ordinary people, often in extraordinary and tragic circumstances. These have been the main source of clues about the nature of the brain. Gradually, those studies led to questions about anatomy and function, some of which had pretty obvious implications.

THE PLASTIC BRAIN

Three million years ago, australopithecines seemed to understand the mortal consequences of a blow to the head, but that didn't require much specialization, as just about any heavy object would do. About ten thousand years ago, Neolithic surgeons found that cutting holes in the head, called trepanation, was a promising treatment for restoring bodily balance, though precise targeting of the burr holes wasn't possible, even if desirable. In one papyrus, the word *brain* is noted six times, and accompanying images depict various head wounds and spinal fluid, along with discussions of the effects of brain lesions on function.[6] The scrolls record that damage to the head affects behavior in many ways. These ancient documents tell us that, even before the appearance of modern *Homo sapiens*, it has long been understood that whatever is in the head is closely associated with life, death, well-being, and even the mind itself.

For more than two thousand years, dreadful injuries were a prime source of intriguing clues about the mind–brain relationship. In the nineteenth century, an iron spike pierced the skull of an American railway worker named Phineas Gage. Though Gage was physically the same person, all who knew the old Phineas agreed that his personality was altered by the accident. Yet, even as recently as 150 years ago, brain science wasn't in a position to make much sense of the details about the changes to Gage's brain that made him seem to be a different person, partly because there were no imaging technologies to visualize Gage's internal wounds.

Today, even an accident of nature, invisible to the human eye, can be observed by modern imaging, as in the rare cases of people born without the brain structure known as the cerebellum, which is responsible for coordinating movement. These people's physical clumsiness can be explained with a computer-assisted tomography (CAT) scan, which shows a large hole at the back of the brain. With this literal insight, investigators have come to understand that some other characteristics also can be attributed to the absent cerebellum—traits like inattention or poor memory, which once might have been written off as simply odd personality traits.

Meanwhile, there were those who had built the road to Delgado's bullring, hoping to tie specific regions far smaller than the cerebellum to specific behaviors and capacities. These "local-izers" emerged throughout Europe and America in the 1800s. Working almost entirely through trial and error, they labored at understanding more specific behaviors and responses by stimu-lating various parts of the brain. Sometimes they were able to learn by observing accidents like the one that befell Phineas Gage.

But there was only so much to be learned from observation. Others, like the Frenchman Jean Pierre Flourens, destroyed parts of animal brains to show that different brain areas were indeed responsible for different functions. Franz Joseph Gall suggested that there were at least twenty-seven specific functions associated with cortical tissue, and he mapped them. But Gall's work was all phenomenological; it was theory unsupported by data. Like many after him, he posited a vast array of brain faculties but no corresponding experiments. And the debate went on and on, about the assignment of specific cortical local-ization of function and in arguments about more distributed and more subtle functions. These discussions feature many, many famous protagonists, including Korbinian Brodmann, Sir David Ferrier, Paul Flechsig, Gustav Fritsch, Franz Joseph Gall, Fried-rich Leopold Goltz, Eduard Hitzig, John Hughlings Jackson, Theodor Meynert, Constantin von Monakow, Hermann Munk, and Charles Scott Sherrington—a veritable who's who of brain science.

As often happens in the history of science when a break-through appears imminent, some took the promise of localiza-tion too far. In the wake of the experiments by Flourens and others, the notion that parts of the brain could be read by com-plex measurements of the skull, a study called phrenology, swept through mid-nineteenth-century Europe and America. Because it seemed to provide insight into both individual and racial differences, phrenology was, for a while, a highly fashionable mode of inquiry, supported by high society and even by royalty.

But, as has been learned repeatedly in the history of brain science, matters were not so simple. Claims and assumptions about localization of function often have been subject to what the logician and philosopher Alfred North Whitehead called "misplaced concreteness," exaggerating the exact location of a function with a particular place. Whitehead's observation was about logic, not brain structures, but it applies especially well to the brain because of that organ's remarkable ability to adapt to new conditions and even to serious injury, a property called plasticity.

The brain's plasticity, its ability to adapt and compensate even when some parts are missing, has perhaps been the biggest puzzle for the localizers. Consider, for example, people who have lost their eyesight. They compensate by expanding their auditory or somatosensory cortical functions beyond the realm of the visual system itself. Although we are a visual species, a blind person can capitalize on (or we might say corticalize) other forms of visualization—a fortunate consequence of plasticity that helps to compensate for an unfortunate event, the loss of eyesight. Or consider deafness. Not surprisingly, compromised auditory function can lead to greater activation of the visual cortex, so that vision is expanded to perhaps capture something of "hearing."[7]

Although there are limits, both hearing and seeing are rooted in compensatory neural capability. Deprive a sighted person of visual stimulation and her brain will compensate with greater activation of the occipital cortex by means of auditory stimuli. Amazingly, Ludwig van Beethoven, who lost his hearing around age thirty, could imagine music sufficiently to continue to compose. He was able to imagine music, a process that correlated with activating his auditory cortex. The intriguing "mirror neurons" in, for example, the primary motor cortex have caused much excitement in neuroscience. It turns out that imagining and doing involve many of the same brain regions.

Still, in some important ways the localizers were on the right track. As more experiences were accumulated and observations about such cases were more accurately reported, it was possible to more systematically assess which parts of the brain were involved with which functions. Although localization of function can be misleading if taken literally, there is no doubt that some capabilities are more closely tied to specific regions of the brain than others.

There also were more cautious, precise, and accurate conclusions. In a landmark contribution to the idea of localization of brain function, another Frenchman, Paul Broca, discovered that an area in front of the brain, now known as Broca's area, is responsible for speech production. This frontal cortical tissue is tied not only to the capacity for speech but also to the ability to structure language, as determined by studies of people with deficits in both capacities. Steeped in knowledge of anatomy, and with the intuition of an experimental scientist, Broca gave localization of function scientific respectability.

Based on keen observation of patients with epilepsy, the British neurologist John Hughlings Jackson confirmed Broca's conclusions about localization. He regarded seizures as electrical discharges from the brain, arguing that epilepsy is the result of unrestrained electrical activity in the brain, a kind of flooding of its circuits. To use Jackson's terms, epilepsy is loss of inhibition in regions of the brain. Others would later suggest that a particular site of origin was an almond-shaped organ called the amygdala. The result is paralysis and loss of control of language, movement, and the tongue. Today we also recognize the role of chemical transmission.

The idea that a certain level of electrical or chemical stimulation, or "kindling," of the brain pushes the discharges past a certain threshold, leading to delusions and convulsions, has been around since Hippocrates. In the amygdala, kindling can enhance startle responses that are associated with a normal event like the fear reflex. The seizures suffered by people with epilepsy might be understood as the pathological extreme of normal neuronal systems function, aberrations of neuronal long-term potentiation. (Long-term potentiation is tied to an idea of the mid-twentieth-century Canadian Donald Hebb, who argued that cells that frequently fire together will eventually tend to facilitate each other.) In the terms of neuroscience, for those patients, electrical firing patterns turn from the appropriate and adaptive to the dysfunctional and pathological and from synchronous firing of networked neurons to nonsynchronous and dangerous firing patterns.

In reaching his conclusions about localization, Jackson was mainly limited to observation of patients who had already suffered trauma. It fell to the American Karl Lashley, originally a behavioral neuropsychologist, to make far more specific observations through ingenious and carefully controlled early experiments with rats. In one series of experiments, Lashley trained rats to perform certain tasks, such as seeking a food reward for a certain behavior. Then he cut certain rigorously identified areas of the rat cortex or in the gray outer covering. The rats with damage in a highly localized section of the brain did indeed have trouble with acquiring and retaining new knowledge, yet they were able to perform other tasks, such as running a maze, just fine.

Interestingly, Lashley saw that this support for localization in one sense also showed its limits, because the lesioned rats were still able to perform in the maze as they had been trained to do. But maze learning might be solvable in many ways. More specific types of learning and memory tasks (such as Pavlovian eyeblink learning, spatial learning, or reversal learning) were later revealed to show more localization of function. Nevertheless, a particular memory isn't located in a single site or "engram," as many had thought, but is distributed throughout the cortex. Modern brain imaging experiments have shown that the distribution of memory is far more uneven and dispersed than Lashley realized.

CHARGED UP

Historically and experimentally, ideas about the role of electricity in the brain are closely linked to ideas about localization. The nineteenth-century Italian scientists (and sometime rivals) Galvani and Volta thought of the firing of brain cells as an electrical event, an idea that implied that behavior could be both excited and inhibited by electricity. Since then, questions about the nature of electrical discharge have fired investigation of the brain's activity and have framed the language of neuroscience itself.

One strategy to determine the localization of function involves pinpointing the sources of electrical discharge, both normal and pathological, and their thresholds. The Scot David Ferrier showed that, in animals, low-intensity stimulation of the cortex (the part of the brain responsible for voluntary movement) could help map which cortical regions are involved with specific motor functions. He also showed that hyperstimulation of those regions could cause a loss of function. Since Hughlings Jackson, there have been ongoing debates about how localized brain functions are, depending on the region, but the general idea that some degree of localization is discernible in the brain's behavioral outputs is not in doubt, in spite of the cautionary tale of phrenology.

Over the past 150 years, electrical stimulation or injury of certain brain regions has been a primary tool in the search for localization of function. Investigations in neuroscience are still deeply embedded in experiments involving electricity, recording activity in cells with a view toward what Galvani called "animal electricity," the electrical currents that flow through muscles and, mostly, within neurons or sets of neurons in specific regions of the brain. These studies led to questions about the kinds of cells that underlie the electrical firing patterns, how new neurons are created, how information is coded into them, and how they transmit across the spaces between nerve cells, called synapses. Delivering electricity to discern function, to isolate some sets of neurons from one another, or to infer primary properties of neuronal tissue all are staple techniques of brain science. Thus, electricity is both a tool to probe and part of a theory to understand.

The twentieth-century American neurosurgeon Wilder Penfield, who worked mostly in Montreal, directly stimulated various regions of the cortex and mapped out possible function by using responses from experimental animals and self-reports from patients. In Penfield's hands, electrical stimulation had also become a medical device. In that tradition, modern brain surgery is able to pinpoint diverse brain regions with a minimally invasive three-dimensional system of coordinates called stereotaxis, so that electrical currents can be precisely targeted. For some illnesses, precision is not required. People suffering from otherwise intractable depression often benefit from electroconvulsive therapy (ECT), which pushes the neural circuits over the threshold and causes a seizure, though the reason ECT can work for these patients is still not understood. ECT is an example of therapeutic neurotechnology without a comprehensive theory.

As in the rest of biology, understanding involves finding patterns, the underlying organizational schemes of living things. Patterns of local and global electrical activity are considered "signatures" of brain cell function, though what those patterns mean remains unclear. We do know that the rhythmic electrical patterns of our nervous system, which work like internal clocks, are diverse and ancient. This includes the well-known circadian system of our roughly twenty-four-hour clock. The nervous system depends on small-scale rhythmic or "oscillating" patterns of neurons that enable large-scale patterns like the sleep–wake cycle. There also are internal generators like the one that stimulates the heart. The internal rhythms of biological creatures are studied by recording these electrical impulses. We now realize that the body has many sources of internal electrical generation, some linked to external events and others running without being tied to them. These coordinated internal generators are linked to joint neuronal firing patterns that underlie the organization of action, coordinating the activities of organs and tissues.

PROBLEMS WITH CATS

In spite of their excessive enthusiasm and theoretical overreach, the localizers did a great service, on the whole, given the information and methods they had at the time. Since then, a major lesson learned from experiments involving electrical stimulation of regions of the brain is about the importance of context: locality depends on various factors. In the 1950s and 1960s, neuroscientists gave special attention to regions thought to be "pleasure centers" and "aggression centers," testing this by implanting electrodes in various parts of animal brains. One classic experiment involved aggression in cats with electrodes placed in their amygdala, found deep in the mammal brain. When exposed to a rodent, these cats expressed predatory behavior. The artificial electrical stimulation reproduced what the cat would have been expected to do in nature. In a second experiment, the cats' sensory field for detecting prey was expanded by electrical stimulation, making it easier for them to detect the prey. In other words, electrical stimulation to the amygdala (or to the hypothalamus) set up the conditions for the cats to detect the prey.[8]

But the results of these experiments didn't turn out to be as specific or predictable as expected when attempts were made to reproduce them. In diverse experiments during the 1960s and 1970s, scientists found that the kind of behavior that resulted from the same electricity delivered to the hypothalamus in diverse species (such as in rodents and in monkeys) depended on the situation. In one context the electrical stimulation would elicit feeding, in another context it led to aggression. Later experiments, in the 1990s, found that stimulation of hypothalamic regions enhanced the salience of food-related events even when the subject did not enjoy the food.[9] Lab animals ate when stimulated, but not for pleasure. Videotapes showed that the animals made faces, as if the food tasted bad, despite the enhanced ingestion.

The stimulated brain regions couldn't be directly correlated to individual behaviors, and they often overlapped with several behaviors. In other words, specificity doesn't mean uniqueness. Various regions produce aggression. Sex can be induced by chemicals like estrogen as well as by electricity. Regions like the hypothalamus and the amygdala have been linked to brain stem sites and to neural circuits. In the long run, over a series of experiments, it became clear that the environment in which the electrical stimulation took place mattered as much as the place-ment of the electrodes in the brain. Just as we've come to appre-ciate that genes in themselves work through proteins and that the gene's expression of proteins is affected by the environment (a field of study called epigenetics), so too have neuroscientists come to understand that brain stimulation can elicit aggression or feeding or other behaviors depending upon the context.

Through the 1950s and 1960s, this research into aggression and feeding was accompanied by the search for the brain's pleasure centers, with similar long-term results. In a famous experiment, certain regions of rats' brains were stimulated, and the rats quickly learned that by pressing a bar themselves they could get an elec-trical charge whenever they wanted, even at the price of starving to death.[10] Perhaps they were experiencing so much pleasure— or compulsive or even addictive behaviors—that they preferred that immediate sensation to eating. But other researchers sug-gested that pleasure might not be the explanation for the rats' seeming preference for the stimulation over survival.[11] Perhaps the electrical stimulation was less about pleasure and more about compulsive behaviors or addiction. When animals are given foods that give them pleasure, they make certain facial expressions, but they do not make these expressions when they eat as a result of electrical stimulation. Nor do humans report pleasure when pleasure centers of their brains are zapped, as might be expected. As one popular song put it, ain't nothing like the real thing.

That said, regions of the orbital frontal cortex can evoke what looks like pleasurable responses when electrically stimulated. But the concept of a reward is more subtle than it might seem. Certain neural circuits underlie an expectation of rewards but, neuro-logically speaking, rewards are not one thing or one brain circuit but many—perhaps as many as there are rewards themselves, in all their diversity. And we need to distinguish (as our brains do) between wanting something and actually enjoying it. It can be hard to remember the difference as we are bombarded daily with advertisements that tickle our wants—called "incentive salience"—for that cigarette or shiny machine or piece of chocolate, even though we may not like the item on offer. Incentive salience is associated with regions of the dorsolateral striatum, a significant feature of addiction. Behaviorists have observed the wanting/liking distinction for a long time in the context of particular foods.

That research has been reinvigorated by Kent Berridge and colleagues at the University of Michigan.[12] Their experiments have shown that certain regions in the neostriatum, mediated by dopamine and endorphin information molecules, are tied to wanting something, and that these can be distinguished from regions connected to actually liking it. Wants are produced, in part, by dopamine-opioid microinjections in that part of the brain, enhancing the motivational allure. When those molecules are blocked, the allure fades, even as the intended object of ingestion is still liked, as in the case of a food that is found to be tasty and desirable but is not obsessively desired. The chemicals tell the story. Dopamine is correlated with increased desire for a piece of chocolate but not with the pleasure of eating the chocolate.

The dopamine effect is quite varied as to object and can be far more cognitively abstract than the thought of something as concrete as a luscious food item. Functional magnetic resonance imaging (fMRI) studies have shown, for instance, that the greater the anticipation of enjoying some music when one is purchasing it, the greater the activation of dopamine and the greater the activation of the nucleus accumbens within the basal ganglia.

All of this leads to a discouraging possibility for the experimenters: maybe the straightforward approach to studying the brain by putting electrodes into brains isn't a model of the way electrical impulses work in nature. Rather, electrical activity is intimately associated with brain chemical production. And, of course, there are risks to excessive electrical currents. Provoking electrical currents artificially in labs can cause all sorts of problems, such as seizures. There is no doubt that electricity matters, but its effect also depends on where, when, and how the charge is delivered and which brain chemicals are associated with it. Theories alone couldn't have told us this electrochemical story. So, again, instead of thinking that theory about the brain should drive what we think brain technologies can do, many neuroscientists now believe that our experience with what the technologies can do should drive our theories about the brain. This is a pragmatic approach par excellence.

CHEMISTRY SETS

In addition to electricity, another way of exciting brain cells came into its own in the twentieth century: chemical secretions. Given a big push in the eighteenth century by Frenchman Antoine Lavoisier's discovery of the role of oxygen in respiration, the conception of chemistry in physiology began to take modern form several centuries ago. By then, many experimenters had come to the conclusion that there must be a control system to regulate basic physiological functions such as heart rate and digestion, processes that couldn't be accounted for by electrical pulses alone. As early as 1877, the German scientist Emil du Bois-Reymond suggested that the brain is aroused or activated by chemical as well as electrical messengers.

The idea of this autonomic, or involuntary, nervous system was tied to the search for chemical messengers known as neurotransmitters, information molecules that are diverse, overlapping,

TABLE 1.1

CLASSIC NEUROTRANSMITTERS IN THE CENTRAL NERVOUS SYSTEM

Catecholamines	Dopamine
	Norepinephrine
	Epinephrine
Indoles	Serotonin
	Melatonin
Cholinergic	Acetylcholine
Amino acids	γ-aminobutyric acid
	Glutamate
	Aspartate

and display divergent evolutionary trends. Ultimately, this idea led to the chemical revolution in neuroscience, to the discovery of a series of neurotransmitters such as norepinephrine and acetylcholine (see table 1.1), and to the appreciation that these chemicals are ancient. Dopamine, for instance, stretches back half a billion years; the monoamines and indolamines like serotonin, so important in modern psychiatry, also are no newcomers to the planet.

One of these old brain chemicals is the corticotropin-releasing hormone (CRH), which is found in many tissues and species. CRH carries information that has to do with the response to fear, to potential danger, or to something unfamiliar, alerting us to internal and external threats. Perceived threats may include just about anything in the environment, from directions in space (the view of the left as "sinister," from the Latin *sinistra*) to spiders to predators, and an adequate response must coordinate a dazzling array of biological systems. So CRH is found in a wide variety of brain regions, including the hypothalmus, the amygdala, and the frontal cortex. Mice that have been genetically engineered to overproduce CRH may be more easily prepared for fear responses.[13] When very young rats are subjected to adverse experiences, they express more CRH as they get into the second week of life, suggesting that a scary environment can create a lifelong foundation of increased sensitivity to perceived threats. If that holds true in humans, it could prove important to understanding the biological source of violence in some children.

Today, neurotransmitters are everywhere in brain science. But the road to accepting the importance of brain chemicals was a rough one that lasted decades and put careers on the line. Like many heated debates in science, locating the definitive answer meant fame and glory for some and frustration for others. As the neuroscientist Elliot Valenstein explains, the notion that nerve cells weren't directly connected took a long time to accept.[14] In the late nineteenth century the top observers disagreed about the facts. Camillo Golgi looked at nerve fibers, using his breakthrough technique of cell staining, and saw a continuous network of nerves. Santiago Ramon y Cajal did the same thing but instead concluded that there were gaps.[15] They reached divergent conclusions from the same data. (The situation was somewhat reminiscent of the distinction between astronomers who, using early telescopes, saw canals on Mars and those who didn't.) What settled the issue was the fact that there was a measurable delay in transmission between the ends of neurons and other neurons or muscles, indicating that there must be a gap between them. Though some still sided with Golgi, by the time C. S. Sherrington described the inferred gaps between neurons as "synapses," in 1897, the matter was pretty much resolved.[16]

Though Cajal turned out to be right, that raised a further puzzle about how those cells could send electrical messages if there were gaps between them. The obvious answer was that there was some sort of chemical process, as theorized by du Bois-Reymond, but how could chemicals cause such a diverse array of events and responses in the nervous system? Over the next decades, an effort ensued that illustrates how science often advances by small and sometimes seemingly unrelated steps.

An early clue was the discovery that one very important heart nerve, the vagus nerve, conveys chemical signals to the brain, and that epinephrine and other chemicals make the heart beat faster. Animal experiments showed that both can speed up and slow down the heart. So that was a kind of proof of principle about nerves and chemicals. But the electrical advocates were unconvinced. Even after two Nobel Prizes were awarded, in 1936, for discoveries about these chemical substances, what Valenstein calls "the war of the soups and the sparks" continued.[17] In the late 1930s, the great Harvard physiologist Walter Cannon elaborated on the concept of "fight or flight" for animal responses to pain or fear.[18] Debates would emerge from the work of the equally distinguished Australian John Eccles, who defended electrical transmission.

Nonetheless, information about the body's chemical production, like that of epinephrine, kept pouring in (no pun intended), and sometimes from surprising sources. The iconic hallucinogen lysergic acid diethylamide, more popularly known as LSD, is famous for its role in 1960s counterculture, but it was also a breakthrough experimental agent in the history of neuroscience. In 1954, the effects of LSD were found to be caused by its effects on serotonin, an ancient and remarkably versatile neurotransmitter that is crucial for digestion, growth, and reproduction. By the 1950s, neurotransmitters were accepted as invaluable to brain science, in both psychology and physiology. Yale psychologist Neil Miller developed an integrated theory of neurotransmitters and behaviors like seeking food, opening up new pathways for experiments with drugs as well as furthering the idea that we might regulate our nervous system by paying attention to its signals.[19]

So the discovery of neurotransmitters encouraged research on specific brain organs known to be affected by certain chemicals, something like the experiments with electrical stimulation. Whereas the electrical stimulants failed to be very specific to behaviors, however, it was thought that the chemical stimulants would be specific. In one 1950s experiment, various animals were injected with sodium chloride at the hypothalamus, a brain structure found in both fish and mammals; as expected, the animals were induced to drink. But in other experiments the chemical stimulants ran up against the problem that the same stuff could cause different reactions. For instance, when male and female rats were given testosterone in one area of the brain, rats of both sexes behaved like females, building nests and rounding up infants. When the same hormone was introduced to a different area, though, they all behaved like males. So hormones, too, turned out to be not completely specific.

Though they're no magic pathway to a full understanding of the brain, the results of chemical stimulation experiments have been encouraging. With much trial and error, the sites to be stimulated can be narrowed down enough to be associated with certain behaviors. The resulting chemical revolution in neuroscience has allowed neurotransmitters such as dopamine, norepinephrine, serotonin, and acetylcholine to be understood in part as modulating general arousal. Norepinephrine may facilitate alertness, dopamine aids in more general cognitive/motor organization, and serotonin helps manage our emotions. With their four-ringed molecular structure, steroids facilitate neuropeptides and neurotransmitters that underlie our attempts to organize our experience, including remembering items needed for survival, such as food, water, and shelter. Broad neural pathways like those for appetite underlie many of the same generic neurotransmitter avenues.

Testosterone is normally thought of as the key male sex hormone, helping in the development of the testes and prostate. But testosterone also is known to play many roles in both sexes, beyond aggression and sex. Many of its effects happen through a process that converts it into estrogen. The estrogen affects both the brain and the regulation of gene products that underlie birdsong, which play a role in territorial defense and aggression. One mechanism that leads to these behaviors is the induction of neuropeptides like vasopressin in brain regions that underlie aggression, such as the amygdala. Besides arousing behavior via the pathways in the brain's limbic system, these chemicals also underlie many forms of motivated behavior as well as, unfortunately, some behavior gone wrong, like the tremors of Parkinson's disease.

Theorizing, discovery, experimentation, and therapy often work together. Take the story of the chemical messenger insulin, which regulates levels of blood sugar. Diabetes is a disease that can cause premature aging of neuronal systems and can affect learning and memory. The nature of memory has proven to be an especially hard problem for brain science to understand and resolve. The localizers were tempted to think of specific brain regions for memory storage, but those efforts gradually gave way to systems thinking that views the brain as a more "plastic" or flexible system in which electrical and chemical systems operate.

Memory is especially important because it enables the life narratives that constitute our personal identity, and its loss though dementias is experienced in large part as a loss of one's very self. Whatever technologies are finally developed to compensate for these terrible diseases will, in the first instance, be thought of as therapies to preserve or restore normal brain function.

These days, many wonder whether such therapies might also help those of us without diseases that undermine cognitive function to become even better than some species-typical norm, to improve our normal memory storage so that overall intellectual function is improved.

Insulin promotes growth in many tissues, including the brain. When insulin production is compromised, the risk of pathologies like depression increases. The increased risk of depression might happen when insulin dysregulation aggravates hippocampal disruption or perhaps glucose compromise in the amygdala. Like many other neurotransmitters, it was theorized before it was discovered—a profound medical science event that led to a Nobel Prize in 1923. In the basic science lab, the discovery of insulin also formed the basis of research on metabolism, especially of carbohydrates and glucose. Taken together, these insights and discoveries and the medical treatments they led to have given normal lives to generations of people with diabetes. With greater understanding of the underlying mechanisms, more improvements for people with diabetes are sure to come.

Whatever medical research and science do to rebalance the human organism through brilliant and hard-won brain technologies—whether drugs, devices, biological materials, or some combination of the three—we are only supplementing the electrical and chemical systems provided by evolutionary nature itself. Unless we understand them, we cannot understand the brain, the rest of the body, or even the way we relate to other people.

Over millions of years, new biological chemical messengers have appeared and diversified, and their genetic codes have been altered and duplicated. Like large animals, chemical messenger species came and went, over millions of years. Some of them, as found in protozoa, bacteria, fungi, plants, and invertebrates, date back to the origins of life. As Charles Darwin noticed, evolution tends to conserve and reuse organ systems in various animals. The same is true of the chemical messengers like endorphins, insulin, thyroxine, estrogen, and dopamine. Molecules like gonadotropin that release growth hormone date back to jawless fish and appear in various guises. Another ancient family of molecules includes oxytocin, sometimes called the "cuddle drug" for its association with certain close encounters. It's pretty clear why biological evolution would want to reuse a hormone like that for sexual reproduction. As it does with other bits and pieces of living things, evolution likes to use the same molecule in different roles. So in addition to sexual pursuit, oxytocin can also be "recruited," as the biologists say, for platonic purposes such as friendship and lactation, depending on what organs express it and where.

Or take vitamin D, another product of a gene that is a molecular ancestor of numerous steroids. Like other steroids, vitamin D binds to receptors that are localized in key brain regions, including the amygdala, the hypothalamus, and the bed nucleus of the stria terminali; brain stem sites such as the parabrachial region and the solitary nucleus; and the spinal cord. Vitamin D is typically anti-inflammatory, tied to neuroprotection and perhaps memory, and it regulates the immune system. A lot of money is spent on supplements by cold-wary consumers, even though there's no evidence that extra doses of D help.

It turns out to be hard to outsmart evolution, let alone to figure out how it works in the real world. Under the heading of epigenetics (changes in what genes do, without changing the underlying DNA), the question of how messenger-producing genes adapt to the changing world in which they function is one of the oldest controversies in biology. Mainstream biology long resisted primitive ideas about how living things acquire new traits (did giraffes get long necks because they stretched to eat?), but as recent evolution-based science has brought molecules into the picture, far more sophisticated theories have emerged (supporting the notion that the giraffes with the growth hormones in the right place survived to produce little giraffes that often had the same hormones).

The precise mechanism for the evolution of gene expression remains an area of intense focus among biologists. One answer almost certainly lies in the ways that genes—and therefore chemical messengers—get turned on and off, called methylation and demethylation. For example, experiments have shown that children with early onset alcoholism have the same rates of this disease whether they are raised by biological parents or by foster parents, suggesting a strong genetic basis.[20] But children raised by foster parents have higher rates of late onset alcoholism than would be expected, perhaps a result of environmental factors like neglect or abuse that reached the mind-brain loop to turn on certain chemical messengers. Epigenetics is no respecter of persons and no promoter of fairness. At the level of the individual brain, Martha Farah has shown a link between childhood poverty and brain development.[21] A large-scale study of more than a million people with the same genomic scores found that those born to high-income fathers were far more likely to graduate from college than those born to low-income fathers.

Neuroscientists are taking hints from nature's methods to craft experiments that enhance or delete chemical messengers. Mice that are designed to have too much of that chemical messenger we talked about before, corticotropin-releasing hormone, can show more fear-related responses than other mice. When CRH is deleted from mice, they still exhibit fear, but not always when they should—not a good thing when predators are around. This result demonstrates that fear involves a number of chemical messengers, including gamma-aminobutyric acid (GABA), which is mainly inhibitory, and dopamine, which is more excitatory. They help organize fear-related behaviors or their pathological forms, anxiety and depression.

We live in an age when chemical messengers and chemical signaling systems can be altered by genetic manipulations. Altering one messenger or system can alter many others. The brain has a basic design that anchors the organism, but it also operates in an environment, a context. Our brains are not in vats, so far as we can tell, and if they were, the vats would be a context to be reckoned with.

2

CONSTRUCTING

Unable to secure a grand theory of the brain in the mid-twentieth century, neuroscientists have used new technologies to reverse engineer the brain, like the precocious ten-year-old who annoys her parents by taking apart that new cell phone. But those parents know that their child's relentless curiosity might lead her to MIT. Or maybe she will learn a bit about how the thing was put together in the first place, about what materials and principles underlie the gadgets she can see through her magnifying glass. There's even a chance that she might come up with some design breakthrough that would only occur to a child.

Perhaps she would do what clever engineers often do with a device they didn't build themselves: as she takes it apart, she figures out how to reverse engineer it, maybe making a better version of the original. Neuroscientists could do the same thing, reverse engineering the brain based on what is known about electrical and chemical signaling systems, the energy requirements and capacities of both, and the spatial and temporal dynamics that are required. The basic design of the vertebrate brain and the various neural designs of invertebrates have been studied for hundreds of years. Some of the most important but also

misleading arguments for evolution have come from embryol-
ogy, often based on images that show that human embryos
resemble fish and chickens during development, or, in the words
of the old saw, that "ontogeny recapitulates phylogeny." Would
that it were so easy.

As the neuroscientist Larry Swanson has observed, we also
know something about the basic design of the brain. After all,
we've been at it for a while. In this information age, we are now
accumulating massive amounts of data about the brain, but that
data only becomes informative when it is put in a context. We
have a wealth of data about brain connections, cell types, and
development, but we have less about what it all means. The field
called neuroinformatics gives meaning to the abundance of data.

Most neuroscientists are a bit more mature than that ten-year-
old taking apart her cell phone, but that once was them, and
even as grown-ups they are no less relentless in their curiosity.
Of course, reverse engineering a living thing like a brain is different from trying to do that with a cell phone, but they do have
certain design principles in common. Among other things, it
would be good if the phone or brain they designed were fast and
no larger than necessary, used energy efficiently, and made very,
very few mistakes, if any at all. The engineers also need to take
into account their gadget's environment, factors like temperature and compatibility with other phones or brains.

When a prototype device like a phone is on the drawing
board, other important questions include what parts it will need,
whether they are already available, and how expensive they are.
If parts are scarce or expensive, the product may simply not be
worth the investment. One of Darwin's great insights was that
living things also have a parts list, that nature conserves and
reuses chemicals and organs. Items might be put on the shelf for
long periods and then taken off as needed. It took a hundred

years after Darwin to appreciate that there is a parts list for living things at a deeper level, as well—the list of DNA sequences made up of amino acids, called the genome.

When sequences are copied by RNA, they can create proteins that are signals for countless functions. Tiny changes in a gene can lead to big changes in the organism. The protein opsin, for instance, is crucial for vision. Changes in just a few of its amino acids out of hundreds make it possible to see red and green. With these manipulations, scientists know how to build an eye that can't do that, as well as one that can, a nice example of reverse engineering. This technique and others have expanded our possibilities for probing the brain.

We want to avoid any confusion that could be caused by our somewhat loose talk about Darwin and design and analogizing living systems to devices like cell phones. It's the same problem that Darwin himself had and that has shadowed his legacy ever since: using common English words to talk about design and engineering does not imply that there was a Designer. On the theological question itself we are neutral, but we are firmly of the view that Darwin himself quite reluctantly came to: the logic of evolution does not require a Designer. Though indeed there might be one, the theory does not require it. Moreover, living things might look intelligently designed in some ways (and we will focus on some examples in the next few pages), but we can also think of myriad ways they could be designed better. Nonetheless, time and the basic components of life forms tend to do that job for us.

Indeed, some biologists think that nature is very nearly programmed to hit on the best possible design. There, too, we disagree. We take a very pragmatic view in thinking about design. One way of understanding our pragmatism is by recalling the old joke about losing something, that no matter where it is, it's always found in the last place you look. If only we could look in the last place first! Like Richard Dawkins and many others, we think that as long as nature hits on something that works for the organism, helping it to survive at least long enough to reproduce and pass on its genes, it usually stops there.

Often, design principles seem to collide. Fewer parts would seem preferable to more. Sometimes fewer parts can work for more functions; other times, they need to specialize. Evolution has "figured out" how to use a single synapse to pass on a couple of different signals between neurons, for example, but to see in both black-and-white and color you need rods as well as cones. There are physical limitations to what you can do with each part, which the engineer has to keep in mind. Ion channels are able to transfer electrical signals, and transporters carry chemicals, but they can't be interchanged. And sometimes there are advantages to having a number of different items for future modification.

All these different parts are hard to understand in themselves, which is why there are tens of thousands of neuroscientists, each of whom is working on merely a few, or even only one, item. No one person will ever know enough to build a brain or will even be able to program a computer to build one—a potentially scary prospect that we take up later. But before we move on to some of the new techniques for building a brain, inspired by knowledge from reverse engineering, there's a prior question.

WHY WOULD YOU WANT TO BUILD A BRAIN?

Some creatures get along perfectly well without a brain. The bacterium called *Escherichia coli* makes do with five types of receptors on its surface. While *E. coli* is making its way through the guts of warm-blooded animals like us, these receptors pick up what it needs to survive, such as glucose or lactose for energy, and forage for more nutrients. To move around, *E. coli* has clever whiplike tails called flagella that snap the bacterium through the muck. *E. coli* even has a simple memory of sorts: the receptors are more-or-less sensitive to their activation for about one second, long enough to let *E. coli* "know" what's going on in its immediate environment, which may affect its survival and growth. It's not much, but this simple system shows that chemistry can do a facsimile of what an encoded memory can do in a brain's information molecule.

E. coli's endowments work well for its size and scale and for its lifetime of about an hour. But, as an individual, it can't adapt to new conditions; it relies on other *E. coli* with the right genetic variation to help the whole population survive. So even if it lived a bit longer, a particular *E. coli* bacterium wouldn't do well if the narrow range of conditions for which it's prepared change. It's also too small for its flagella to get it far, so if things do change, it can't forage beyond a limited space—less than a nanometer, or 1 billionth of a meter. What the philosopher Thomas Hobbes said about human life in the state of nature applies a million times over to the life of *E. coli*: it is nasty, brutish, and short.

The one-celled creature called a paramecium is a big step up. Guided by its receptors through ponds where it forages widely, it feasts on luscious rotted vegetation and sludge. The paramecium is several hundred thousand times bigger than an *E. coli* bacterium and has hairlike bodies called cilia that sweep along its sides for motion; it is even able to reverse course and sling itself around obstacles. Paramecium's agility is made possible partly through an important innovation: calcium ion channels that open and close when its head end bumps up against something, signaling the cilia at the tail to engage in a back maneuver. On the whole, it's not a bad life for a single cell. Still, what works for a one-celled organism won't work for a creature with many cells. For that, a brain would help.

Boasting a primitive brain, the thousand-cell *C. elegans* is a worm with a diet not unlike that of paramecium, but its crawl allows it to range around a great deal farther for dinner, and it has a digestive tract, along with brain chemicals that make it forage for food. It also has a sex life, facilitated by its ability to travel through soil, and releases and senses a pheromone, a chemical that facilitates socializing.

In support of grazing for food and mates, *C. elegans* is equipped with a brain that takes up a bit less than half of all its cells, suggesting how much mentality is required to keep even such a simple beast in the game of life. That four-hundred-cell brain includes seventy-five motor neurons that coordinate its ninety-five muscle cells, enabling it to achieve very precise but highly adaptable contractions, depending on the terrain and objects it encounters. However, to spare the brain the need to organize the specific patterns of movements involved in crawling around, there is an electromechanical feedback loop between the muscle and motor neurons. By distributing the information system throughout the worm's body, the brain doesn't have to use so much energy. The worm body itself is a kind of distributed computer, a simple illustration of a concept that is exploited by far larger and more complex creatures, including us. As we've said, the brain isn't only in the head.

SENSING AND RESPONDING TO THE EXTERNAL WORLD

E. coli, paramecium, and *C. elegans* only have as much memory as is needed for survival and reproduction. But once a creature is large and mobile enough to get beyond pond scum and soil, it needs a lot more neural infrastructure. What might loosely be called memory includes registering both external and internal information and the ability to process it into patterns that select for certain behaviors, like eating or fleeing. Making sense of the external world seems like an obvious requirement, but a bigger brain is also needed to monitor its internal milieu and to coordinate what's out there with what's in here. Managing context means matching response to environment.

All kinds of specialized cells, tissues, and organs are required to accomplish that matching. Although we're accustomed to thinking that internal processes work largely on their own, they are in fact in deep contact with the world around them. Real autonomy would mean death. For example, when the brain senses patterns that match up with signals for food, the digestive system prepares for what might be coming, starting with salivation. At the other end of feeding, as the stomach's contents are excreted, the body readies itself for more foraging. The wondrous zebra fish changes color to mask exposure to a potential predator, to fade into the landscape. Information molecules like the corticotropin-releasing hormone discussed earlier facilitate the change in the fish's pigmentation. The same molecules in its brain facilitate behaviors that avoid light. The expansion of information molecules takes place across diverse forms of behavioral adaptation.

The components of any creature's brain follow several design principles found in living things more generally:

- Make each component irreducibly small.
- Combine irreducible components.
- Compute with chemistry whenever possible.
- For speed across distance, compute electrically and minimize wiring costs.

Here are some applications of these design principles: Keep the number of brain cells to a minimum and keep them a certain thickness. Their membranes need to be not too big, not too small, but just right to be efficient and permeable and yet to hold together. Brains need to be big enough to do what the organism needs yet must fit within a skull, and if more surface area is needed than the skull can afford, brains can fold in on themselves, as they do in primates.

In its internal communications, the system favors changes in electrical potentials—the charges required to carry energy—rather than chemical pulses, which are slower and take up more of the brain's resources. That preference for electrical charges to convey information also means that as much work as possible is done within a cell rather than between cells, which requires the more costly and cumbersome chemical, rather than electrical, messengers.

When messages are being sent, economize by sending only what is needed, send at the lowest acceptable rate of speed, and minimize the length and diameter of the "wires" that are made up by groups of brain cells. The rules that underlie the transmission of information are critically related to survival. As Peter Sterling and Simon Laughlin elegantly express it, "Information is the reduction of uncertainty about some situation X associated with any variable Y that is causally correlated with X."[1] Survival requires that the creature encounter the fewest possible surprises in the course of its routine.

The system anticipates, responds, and resolves. It responds to behavioral drivers and motivation, to incentives to eat, move, and reproduce. These incentives are internal as well as external. Chemicals like dopamine are among the internal incentives. They organize behavior partly by providing rewards when the conditions for their release are satisfied. With each activity, internal functions like blood pressure and kidney volume have to be prepared and adjusted. In a complicated organism, well above the worm, many activities require preparation and resolution in light of the context—socializing, sleep, sex, and on and on.

Bigger brains that maintain a stable internal milieu and suitable responses framed in terms of its needs and available resources enable wider behavioral repertoires in a creature that travels widely. The external milieu includes the passage of time, so by the time we get to the worm, living things are loaded with oscillators (even plants have estrogen), both in the brain and distributed. These internal clocks, which register an approximate twenty-four-hour day, are another key in neural design. The organism makes an impressive symphony of decisions in real time.

But there are also constraints. Brain cells can only fire so rapidly, can only process so much information at a reasonable energy cost. Bigger brain design takes inherent limitations like these into account, partly by being organized in certain ways—like on the scale of fifty thousand neurons per cubic millimeter. That's a lot.

SYNBIO

With these design principles in mind, the most obvious way to build a brain is to use as raw materials the chemicals found in living organisms themselves, and then to set about reproducing their processes and structures according to what we know about the ways in which they are engineered in nature. That's what the fairly new field of synthetic biology, or "synbio," tries to do, using a tool kit adapted from living things. The most intuitively sensible way to do this is from the bottom up, using nothing but chemicals. The other approach is top down, reconfiguring existing biological materials according to the goals of a particular experiment.

The roots of synbio go back to at least 1972, when DNA was spliced, or recombined, from a bacterial virus to a monkey virus. Recombining DNA in lab animals has come to be a fantastically useful technology that, among other things, has enabled scientists to understand what individual genes do under various environing circumstances, a field called epigenetics.

About ten years later, another big development was a technique called polymerase chain reaction, which copies and enlarges sections of DNA so that they can be more easily managed in experiments. After another decade or so, machines had been developed that could quickly determine the segments of genes that make up DNA. Taking advantage of these gene sequencers, a vast quantity of information flowed from the Human Genome Project, making it possible to identify the genetic codes of organisms, from bacteria to human beings. Then scientists began to synthesize those pieces of DNA to make new sequences. In principle, a complete organism could be created by synthesizing DNA fragments to create a new genome. The J. Craig Venter Institute reported, in 2010, that they had synthesized the genome of a bacterial cell and transferred it into a cell that had lost its genome.[2] The cell then replicated itself according to the instructions in its new genome.

Did the Venter team create life? It depends what you mean. They argued that they did, according to the criterion of reproduction under the command of the genome, but others argued that there had to be a host cell that made that possible. We're not going to settle that argument here. At least Venter's group did establish a proof of principle that the tool kit of synthetic biology can create a novel organism that can function like one in nature. And that's a pretty big accomplishment, by any measure. Another step in this story is being undertaken by groups like that led by George Church, in a project called GP-write, which aims to build large genomes and test them in cells to see what various combinations of genes can do.

Synbio could, in theory, scale up to much bigger brain power than is needed by that synthetic bacterium. But "in theory" is a lot easier to say than it is to prove, and until the genetic components and actual functions of various brains are better understood, it's not clear what you'd be trying to build. Instead, neuroscientists study the brain by using computer simulations. Admittedly, in silico seems a far cry from in vivo, but given the right biological information, it's not as far as one might think. As it happens, computing power played an indispensable role in the series of discoveries that led to synbio, embodied in the genetic sequencers themselves.

COMPUTING THE BRAIN

Rapidly growing computing power and speed has transformed both genetics and neuroscience. So-called supercomputers are crucial tools for creating models of living systems so that scientists have an idea how a system will behave, sometimes even before it's possible to actually build it. At the same time, more information about how one or a handful of cells work helps to create more useful simulations. So computer simulation of neural systems is a two-way street between designing the simulation and learning about the actual cells.

An impressive example of the way this process can work was accomplished by the Blue Brain Project, in 2015, after a decade of work.[3] The goal was to build a computer simulation of the brain by using information about the three-dimensional shapes of neurons, the proteins they produce and the channels they use, and how they communicate with one another through electrical impulses and various channels. The announced result was a digital reconstruction of thirty-seven thousand neurons, a portion of a rat brain about the size of a grain of sand. At the very least, this effort yielded an impressive display of information about neural activity and brain architectures.

The Blue Brain Project also showed how expensive brain simulation can be. This representation of only a tiny portion of a brain required the labor of dozens of scientists and the investment of enormous resources. And there's still a lot of biology missing from this partial draft of a minuscule portion of a brain. Blood vessels and many other features were not included. Its defenders say the project is only a first draft and that more big computers will be unleashed to improve it, but its critics say that there are less complicated ways to accomplish the same goal and that a complete digital reproduction of the brain is not necessary to make a simulation that can teach about brain function.

Whoever turns out to be right about that, there's no denying that the rat brain circuit study proved to be a remarkable platform for displaying a lot of information about the anatomy and physiology of neurons and of the synapses between neurons. Like so much of modern neuroscience, that platform relies on the insights of a late-nineteenth/early twentieth-century Spanish scientist whom we mentioned earlier, Ramon y Cajal. Appropriately enough, the Spanish part of the Blue Brain Project is named after him. Cajal's staining method allowed him to visualize structures that couldn't be seen with the microscopes of the day, including spines between neurons, which were extensively studied with electron microscopes more than half a century later. (Cajal's staining method produced such beautiful images that, in 2018, they were exhibited at a New York University art gallery. Not long after Cajal, the philosopher John Dewey noted that an artistic sensibility runs through the veins of scientific inquiry, as indeed was the case with Cajal's drawings.) Some structures are still hypothetical but remain useful concepts. The idea that cortical neurons are organized into "columns" that respond to the same stimulus has been seen by some as a key hypothetical element in neural design, though not everyone agrees.

In neuroscience, brain-building takes place in computers. That might seem like a bit of a head snap (so to speak), but there are some eminently practical reasons for doing things this way. For one, simulations help guide experiments by generating hypotheses that can be tested. The simulations might be only partial and may leave out many factors, as the Spanish Blue Brain Project left out blood vessels, but that doesn't mean they can't provide useful information. Still, unlike simulations in physics and genomics, brain science simulations will often have the problem that there are so many different levels of processes that might be left out, and this could have a serious impact on the meaningfulness of the results.

At a still more practical level, researchers must contend with the complexity of the way modern brain science is done. Since there is so much information being gathered about the brain, different aspects have to be farmed out to different labs in many countries. Those many teams, working on so many different submodels, can easily fail to coordinate their understanding of the assumptions they're making, including the implications of previous research. There's also the danger (present in all sciences that work on simulations, such as climate modeling) that conflicts in the models will be "smoothed out" to accommodate unexpected or conflicting results.

Modern neuroscience projects at the level of the Blue Brain or Human Brain Projects are expensive and involve public investment, often from a number of countries. Understandably, those publics expect that the work will be transparent. With hundreds of millions of dollars involved, some of the decisions neuroscientists make about how to approach their work, like which previous projects look like the most fruitful path to a next step, are matters about which reasonable people may disagree. These are problems of resource allocation that might puzzle non-scientists. And scientists themselves can and do disagree about, for example, whether large-scale, centrally managed efforts like the Human Brain Project, for instance, are as cost-effective as smaller, "investigator-driven" studies.

Moreover, the highly public nature of these big neuroscience projects doesn't always come with clear explanations or understanding of the meaning of the tools being used. In any science, neuroscience included, simulations are tools to develop and (one hopes) to answer very specific and largely hypothetical questions raised by the simulation itself. But in the highly public world of big science, it's easy for the simulations to be mistaken for descriptions of actual brain processes. As often as not, that is a misconception, though an understandable one. Somehow, the message has to be gotten across that simulations are not reproductions. British writer and critic Rebecca West, discussing works of art, wrote, "A copy of the universe is not what is required of art; one of the damned thing is ample." It would seem that this also applies to simulations in science.

WET LABS FOR BRAIN SCIENCE

The brain is hard to study in context. For many experimental purposes, it might not be necessary, and certainly is not convenient, to build a whole brain, let alone one that is put into a living thing. Brain functions such as learning and memory would be far more productively studied with a more accessible and flexible laboratory model than has been available. Nonetheless, the developing human brain can't be accessed in utero, and there are well-known limits to animal brain models.

That explains the excitement about two new "wet lab" systems that enable neuroscientists to experiment in ways that formerly were impossible. One such system was officially announced in 2019, in the journal *Nature* and with appropriate fanfare in the popular press. A Yale team led by Nenad Sestan had developed an artificial blood that enabled them to reestablish some basic cell functions in the brains of pigs that had been dead for four hours. This was a result that surprised the investigators themselves, not to mention the larger community of brain scientists who were not part of the experiment. Summarized in their abstract, the writers noted, "We observed preservation of cytoarchitecture; attenuation of cell death; and restoration of vascular dilatory and glial inflammatory responses, spontaneous synaptic activity, and active cerebral metabolism in the absence of global electrocorticographic activity."[4] Sestan and colleagues were careful to follow ethics protocols and, to ensure that nothing resembling higher functions, "consciousness," or any sensation at all would be experienced by the subjects, they surgically isolated the brains of the thirty-two animals, with various blocks and plans to cool the brains if any such evidence emerged. Nonetheless, the results demonstrated basic mammalian brain reanimation, presenting a long-term hope for neurorehabilitation to ameliorate the symptoms of conditions such as stroke.

Cerebral organoids are systems of cells that mimic some of the attributes of a brain. What's remarkable about brain organoids is that, in the right environment, the cells self-organize into structures like cortical layers and produce human-typical cell types and progenitor zones. Organoids can be made for just about any organ, thanks to progress in stem cell biology, but they are especially intriguing for learning more about neurodevelopment and brain functions, especially with regard to the outer radial glia, which have been largely inaccessible to direct study. Organoids have been studied for factors associated with microcephaly and to model disorders such as schizophrenia that exhibit abnormalities in myelin and interaction between neuronal and glial cells. Synaptic pruning, the normal developmental process that ends at puberty, goes awry in many people with psychiatric disorders and could be modeled in organoids. Sometimes called "mini brains," organoids don't have the same anatomy and architecture as a brain, which isn't surprising, since they don't get much bigger than a lentil. However, they are three-dimensional, which is a great advantage over two-dimensional cell culture models, and they can survive indefinitely, though they reach maximum size in two months. As more labs adopt organoid models, the models' architecture also needs to be standardized for experimental reproducibility.

Brain organoids present a good opportunity for our first explicit mention of ethical issues in neuroscience, or "neuroethics." Should people who are donating their cells be asked for specific consent to create organoids from their cells? People with schizophrenia who suffer from delusions may be intellectually capable of giving consent, but the idea that portions of their brains will be grown in a lab is enough to make anyone anxious. What about children with diseases like autism? Presumably, if any permission is required from minors, it would be asked of their parents or legal guardians. Could the children overrule that permission when they attain the capacity to decide for themselves whether they want their cells to be in an experiment?

Still more disconcerting is the remote possibility that, if grown long enough, into complex structures, the organoids could develop something like perceptions. At a more practical level, human cerebral organoids have been shown to integrate with the brains of rats. There is a widespread scientific consensus that human neural stem cells shouldn't be put into the brains of other animals. Should that rule include organoids, which many believe show far more promise than do any animal models for important research with pluripotent human stem cells? That ethical question remains to be resolved.

In the end, though, no matter how far brain organoids develop, whether in vivo in other animals or in vitro in lab dishes, we think the most they can do is to help neuroscientists understand some basic mechanisms of diseases that originate in the brain. That is reason enough to pursue basic science questions and will perhaps lead to much-needed new treatments. But the knowledge gained from organoids is limited by the fact that they are not brains in the rich and challenging context in which brains become all that they can be.

SEARCHING BRAINS

The twentieth-century American psychologist Karl Lashley spent much of his brilliant career in search of the engram, a hypothetical location in the brain where memories are stored. He was frustrated in the search because much of what we call memory is not located in a particular place. That much is clear. We would go so far as to say that not even the brain alone contains memories, whether localized or distributed. Rather, as the philosopher Andy Clark has said, we routinely scaffold onto the information within the environment to which we are adapting or trying to adapt.[5] After all, we are part of the environment and the environment is in us. This is what we might call the extended or expanded mind. Our innate capability to bootstrap easily to resources in the environment couples with diverse learning skills that underlie the human brain. In effect, we upload memory storage in familiar environments.

Our tools for investigation expand our basic abilities. We investigate the world around us with others because problem solving is a social phenomenon, with science itself as the most intensified example. The brain is active, exploratory, and also conservative, because neural resources are metabolically expensive. The brain already consumes so much of our caloric intake that it makes sense for the system to presume certain expected outcomes, to form habits. Prediction usually gets us where we want to go because of the accumulation of previous experience. It maximizes resources, and thus neural mechanisms are tied to prediction.

In an approach reminiscent of the pragmatists Charles Peirce and William James, Karl Friston and colleagues have talked about the minimization of surprise and, when predictions are erroneous, the updating of the system, as in the "deep learning" of an artificial intelligence (AI) system.[6] Habits are in this sense codifications that help to reduce the expenditure of neural energy. (Genes like FOXP2 underlie the motor control that enables habit formation.) Emotions or affect also play an important role in this conservation of energy by means of prediction. The brain is laden with sensory motor integration across the nervous system. It is an active device set in the context of problem solving, and appetitive (search) and consummatory (satisfaction) behaviors prefigure much of what we do.

The brain is always active, whether in a resting or a default state with no task, or when it is task-oriented. Certain regions of the brain are essential for what has come to be characterized as the predictive brain, where prior probabilities are computed. Bayes' theorem, named after the eighteenth-century mathematician, provides one possible explanation for how the brain makes reasoned use of uncertainty in the course of integrating information and making predictions. The probability that certain learned events will recur is taken into account by the brain as it organizes our experiences. The appetitive (search) phase is aided by the calculator's capabilities within normal function and by habits of action that minimize excessive consumption or energy exhaustion by the brain. We know that the search phase is metabolically very expensive, a glucose-consuming neural machine that is not well adapted for the long term. Glucocorticoids are anti-inflammatories that restrain the effects on neural tissue, essential for their long-term health.

Many forms of learning are activated when habits are disrupted by unfulfilled expectations. The brain then keys into a search mechanism, which involves more learning and more energy consumption. In other words, the strategy is to maximize efficiency whenever possible, often by enhancing predictive capability and thereby reducing neural energy consumption. Many forms of organized syntactic relationships, like language and movement, are organized in the motor cortex and basal ganglia. Habit formation is essential to the organization of action. Steroids fortify the binding of neural sites, including binding corticosteroids on basal ganglia, and recruiting their resources to stabilize and normalize neural function.

CONSCIOUSNESS EXPLAINED. MAYBE.

William James famously and rightly suggested that we should think of consciousness less as a thing and more as a range of events with focal points in our awareness as we cope, or fail to cope, with our environment. Consciousness is rooted in attention to events that affect us and that in some way matter to us. The active mind (better called mind-*ing* than mind) is engaged with a range of events for which the individual is present. These experiencing events are rooted in biology—the biology of adaptation. We may not be able to define whether an event is conscious or not, but that is less important than granting that there is a range of ambiguities in the way events come to our consciousness, in the way they become part of our experience. For us, as we are conscious of being in the world, we are oriented toward events that are occurring, events in which we may or may not participate.

James stresses the vast range of events that are associated with our consciousness, noting that our personal identity and sense of self are tied to memory, and that memory and attention are features of his celebrated notion of the "stream of consciousness." So far at least, computers don't have it, and by definition zombies never will. Can consciousness be reproduced in some combination of organic and inorganic material? Life scientists tend to believe that, even in such a hybrid mind, consciousness will still be lodged in the organic or biological contribution. Physicists aren't so sure, noting that molecules, the basic components of life, are also the components of everything else, so why not of a nonbiological mind?

Suppose an "abiotic" mind is possible. In that case, consciousness probably will not be created simply through a brute force combination of discrete elements, like the digital capacity of a supercomputer, but as a stream. Biology cannot be reduced to a list of functions. At least, that's what appears to be true based on what is known about our animal nature. As James asserts, "The existence of the stream is the primal fact, the nature and the origin of its form the essential problem of our science."[7] Is what David Chalmers calls the "hard problem of consciousness"—the relationship between a physical object (e.g., a biological system called the brain) and the quality of being conscious (i.e., much of what the brain does)—a real problem?[8] Or is it a pseudo problem, an apparently meaningful idea masquerading as a problem? One thing is sure: consciousness involves something about what it is like to be "on the inside," something like Descartes's cogito perhaps, or what the philosopher Thomas Nagel captured in the title of his famous paper "What Is it Like to Be a Bat?."[9] Whatever it is like to be a bat, it is like something, some quality of experience, something on the "inside."

Consider, for example, the challenge of simulating the brain's information processing and, as many would like to do, speeding it up. Neuroscientists search for suitable material to aid the brain or to simulate and expand neural capability. But there should be no confusion between simulation in artificial material or non-biological tissue and stimulation in neural tissue. It is not clear that increasing neural computational capabilities will make them faster than silicon or other "artificial" materials for information processing. Numerous kinds of material that can achieve enhanced neural computation are being studied. Although it is difficult to put such material into brains, we believe they ultimately will be transplanted—when the problem of the body rejecting nonbiological tissue is resolved.

Indeed, rejection of nonbiological tissue is a reminder that the brain is a living thing and not simply an inert computation device. Wet tissue chemical signaling systems have to be understood and integrated within the nonbiological tissue. For instance, the immune system's signaling processes are in part a reflection of the activation and regulation of information molecules such as cytokines. Most of the immunological chemical messenger systems, on the other hand, are peptides or neuropeptides. Understanding how to integrate these signaling peptides is essential for the acceptance of nonbiological material by a noncompromised immune system.

Thus, learning about rejection and acceptance of artificial tissue within the biology of the brain represents a fusion of the strategic development of simulation and reproduction in the characterization of brains. For now, we can sidestep the issue of how exact the simulation of brain activity will be and just note that the ability to simulate would go a long way toward understanding the brain's information processing. At that point, we would no longer have to put artificial tissue and brain tissue in two different categories.

A true brain simulation will have to be more than merely functional. "Just try—in a real case—to doubt someone else's fear or pain." This was Ludwig Wittgenstein's challenge to the fearsome problem of "other minds." Here Wittgenstein asks us to go beyond Cartesian doubt, which is always possible but trivial (in the sense that it is not a self-contradiction to formulate a proposition that doubts that someone else is conscious), to what we would call pragmatic doubt.[10]

That sort of doubt is not trivial. Psychopaths are those who are unable, "in a real case," to appreciate that someone else is anything but an object. In a normal brain, the experience of someone else's fear or pain excites deep and ancient processes of empathy, which requires some concern for others. Regions of the neocortex are active during the sense of empathy—the interpretation of another's misfortune and of their historical and personal circumstances. Perceiving another's misfortune, their psychic or literal pain, requires a wide array of both cortical and subcortical tissue. Empathy entails taking the time to consider the context of another, to move a little beyond oneself. It is easy for some, not for others.

Philosophical doubt about other minds isn't a viable excuse for sadistic behavior. Since ancient Greece and Rome, the legal system hasn't granted even psychopaths a free pass. Violent crimes are crimes, end of story, and courts of law determine guilt or innocence, also end of story. However, depending on extenuating circumstances, the most harsh sentence, such as the death penalty, might be waived if, for example, the offender her- or himself has been a victim of abuse and exhibits some neurological disorder.

Nor do philosophical doubts about other minds let us off the ethical hook in extraordinary moments, such as when it must be determined whether life supports are to be withdrawn or death is to be declared. The persistent vegetative state is a condition in which there is no awareness of the world or of the self. This condition has become well known due to legal cases like that of Karen Ann Quinlan, in the mid-1970s. Quinlan was stricken at a time in the history of medicine when excellent care could provide artificial feeding and manage her infections. She survived for more than nine years, with enough brain stem cells firing to enable her to breathe on her own but still in a persistent vegetative state. It is now generally agreed that if the patient would not have wanted to live in this state, artificial feeding may be withheld at the request of a legal decision maker acting on the patient's behalf.

New evidence from neurological exams suggests that some patients might be in minimally conscious states, rather than in persistent vegetative states, so that they do have some level of awareness and are therefore not vegetative. At some point, however, virtually all of these patients progress to what is known as brain death, which has come to substitute for the traditional standard of cardiopulmonary death. In the decades since a 1968 ad hoc Harvard committee report, every U.S. state has adopted some version of the Uniform Definition of Death Act, which is based on that Harvard report (though New Jersey has a "conscientious objection" to brain death on the books), but the technical means of reaching a brain death diagnosis remain somewhat varied. Perhaps even more contentious are debates about the beginning of awareness in the human fetus, especially with respect to the ability to feel pain. The very fact that these medical ethics controversies remain so vigorous and emotional is a demonstration that, despite metaphysical debates, consciousness is taken very seriously indeed.

SIMULATING BRAINS

Setting aside hard problems like the existence or nonexistence of other minds, can a conscious brain be simulated? Neuroscientists have largely avoided the question of the nature of consciousness. They mostly consider it inscrutable and are happy to give it over to the eager philosophers who dabble in such obscurities. However, the idea that, with the right design principles, brain simulation is possible is one of the factors that is making such conversation a bit more respectable

It was the genius Alan Turing who, in 1950, proposed the famous test that bears his name, in which a computer's responses to a series of questions could render it as good as conscious. (There's been a lot of debate about whether the test would "prove" that the computer is conscious or merely that a human inquisitor has been gullible enough to take it for conscious.) Drew McDermott is a computer scientist who has developed an interesting account of the ways in which the field of artificial intelligence can inform the question of the nature of consciousness, and especially the question of whether there will ever be a conscious computer. In a review article, McDermott begins with a straw person he calls the Moore/Turing inevitability argument, which is itself a reconstruction of an argument by Hans Moravec and Ray Kurzweil.[11] This argument combines the Turing Test with Moore's Law, which forecasts a continued doubling in computing power roughly every two years, resulting in the following quasi syllogism, as expressed by McDermott:

1. Computers are getting more and more powerful.
2. This growing power allows computers to perform tasks that would have been considered infeasible just a few years ago. It is reasonable to suppose, therefore, that many things we think of as infeasible will eventually be done by computers.
3. Pick a set of abilities that, if a system had them, we would deal with as a person. This includes things like having a normal conversation or being a talking robot that can play poker well. We would treat it just like a person who sat at the same poker table.
4. We would feel an overwhelming impulse to attribute consciousness to such a robot.
5. The kind of overwhelming impulse to treat the robot as conscious is the only evidence we have that a creature is conscious. It is, in fact, the only evidence we have that real people are conscious.

However, McDermott himself considers the Moore/Turing inevitability argument to be question begging, because it assumes what it aims to prove: that there will be an "overwhelming impulse" and "overwhelming evidence" that the computer is conscious, according to the same criteria that we apply to other humans. His own view turns on arguments and evidence that consciousness is a result of intelligent systems' modeling of their own processes. After we move our arm to pick up a cup of coffee, our system accounts for that behavior by self-modeling a consciousness that represents it as an intentional action. The modeling is vastly more efficient than the system that explains all the mechanical and chemical processes that led to reaching for the coffee cup. Any intelligent robot that was forced to deal with its presence in the world as we are would have the same kind of modeling capacity that we do. Consciousness is a self-simulation.

We can't reproduce all of McDermott's argument here. What we find telling is that the philosopher Daniel Dennett seems to come remarkably close to McDermott's view. A longtime skeptic about the accuracy of introspection, Dennett analogizes consciousness to the manipulation of file icons on a personal computer desktop.[12] We don't need to know all about the ones and zeros and the microchips that are doing the actual work; the file icons are enough. In fact, if we did have all the details of the activity of the software and the hardware at every moment, that would be an insane form of information overload. Instead, evolution has equipped us with a convenient "user illusion" called consciousness that is beautifully fitted to our needs. This is not a question we can pursue in detail here, other than to note that the neuroscience of brain simulation is influencing philosophical accounts of the nature of consciousness.

The philosopher Nick Bostrom has argued that, compared to a biological construct, silicon-based machine intelligence is a more plausible, even inevitable, route to the construction of something that is the functional equivalent of a biological brain, but that this also is a potentially quite dangerous route. A machine that is as intelligent as a human brain could both program itself and develop other machines that it could integrate into its system, thereby vastly expanding its computational capacity, to the point that it would achieve what Bostrom calls superintelligence.[13] Suppose such a device were to develop certain goals that would serve the completion of its computational task (e.g., the solution of a seemingly impossible mathematical problem). In that case, it could, in principle, subjugate every bit of matter on Earth—and perhaps beyond—to the job of information processing. Such an outcome not only would mean that human beings would be entirely dependent on the superintelligence for their survival but also could lead to the end of human life itself.

Bostrom's ideas have generated a lot of controversy, but he is not alone. The late cosmologist Stephen Hawking and the entrepreneur Elon Musk are among the notables who have also expressed alarm about the possibility of a machine intelligence takeover, while others, like the philosopher John Searle, think that Bostrom's whole argument is misconceived. Rather than settle that question, we are mainly interested in the idea that the seat of humanlike intelligence need not be based on biology. In principle, there doesn't seem to be any reason that intelligence can't be silicon-based, though there's a long, long way to go before computers can get there. Quite apart from building a brain or its functional equivalent, the kinds of "brain apps" we are talking about in this book are surely harder to design and implement than modifying a computer program. As Bostrom notes, the basic methods of enhancing biological intelligence—sleep, nutrition, exercise, avoiding disease—only take you so far. The benefits of neurobiology and genetics are mostly off in the future. But perhaps not so far off. Gadgets like cell phones and search engines help, especially when combined with older collective intelligence systems like seminars and professional journals, but getting people smarter in these ways is really grinding it out.

But Bostrom doesn't want to abandon neuroscience. On the contrary, to get to the point where the advantages of machines could be fully exploited (like creating better learning rules than are possessed by biological brains), their architectures should be modeled on the human brain. That requires the creation of three-dimensional maps based on extremely thin brain slices and computer-generated images, the kind of work being done at the Allen Institute for Brain Science. Founded in 2003 with hundreds of millions of dollars from the late philanthropist Paul G. Allen, the institute's hundreds of scientists spent their first decade acquiring massive amounts of data that enabled them to describe brain cells and their development in rodents, macaque monkeys, and humans. A big part of the next phase of their work, which they call their 2020 vision, is based on the cerebral cortex of the mouse, which boasts the same basic hardware as that of the human brain. This high-risk idea is to chart the anatomy of the cortex, the brain's "gray matter," and then measure its electrical activity to learn how its structure determines its function. The mouse brain–based computer models that result from years of work will, it is hoped, include brain observatories or "mindscopes" that will be part of the equipment of neuroscience labs all over the world.

The Swiss neuroscientist who heads Europe's Human Brain Project, Henry Markram, is in some ways Bostrom's opposite number. Far from worrying about the existential risks of machine superintelligence, Markram wants to use the billion euros he convinced the European Union to invest to simulate human intelligence in a computer. Closely associated with the Allen Institute mission, the Human Brain Project's strategy is to crunch the massive quantities of data coming out of neuroscience labs all over the world, using new computing platforms that will eventually combine all that information in a single system. Those platforms will include new hardware like improved computer chips and new mathematical models of brain activity. Pretty much everyone agrees that more investment in organizing the data deluge of neuroscience is a good idea that could lead to great innovations in just about every field, including medicine.

Yet not everyone agrees that human brain simulation is even possible, let alone desirable. No one really knows what it would even mean to simulate a human brain, because simulation isn't reproduction. For example, Markram's conception of the Human Brain Project was revamped after cognitive neuroscientists noted that there didn't seem to be room for research on higher brain functions like thinking. To them, all the project can do is patch together a lot of information about how brain cells are hooked up; it cannot reveal how real brains are able to do what is distinctive about them, like learning. If they're right, then rather than worrying about the future of human beings, neuroscientists should be worried about annoying the European Union with a lot of overpromising.

It might be that people like Bostrom and Markram are talking past each other. Bostrom and others who take his approach seem largely uninterested in the biology of thought. Like their predecessors, the functionalists of the 1970s, they seem to believe that the material itself is immaterial. Markram and his allies believe that the chemistry matters, so that function falls out of displayable information. Yet they share an underlying assumption, common to many, that the brain can best be understood as a computing machine. That metaphor for the brain, like so many that have gone before it—humors, hydraulic engines, animal spirits, clocks, electrical generators, and so on—has severe limitations, of which we should be wary. All of these metaphors help us organize some of our data about the brain, but none capture all of it, or its context.

SMART MICE OR SMART ARTIFICIAL INTELLIGENCE?

Whatever the prospects for brain simulation, we think there's another, overlooked issue with the way neuroscience is being leveraged to make AI systems smarter. The prospects for making rodents smarter with implanted human neurons have dimmed, but the prospects of a smart computer continue to grow. In 2016, a computer program called AlphaGo showed that it could defeat a professional Go player.[14] Those machine learning algorithms continue to teach themselves new, humanlike skills, such as facial recognition. Except, of course, they are better at it than the typical human. By contrast, another paper indicated that implanting systems of human neural cells called brain organoids in mice didn't make the mice smarter maze runners.[15] There's a lot of regulation and forehead furrowing about the rodents, and plenty of freaking out about long-term "existential risks" of AI, but not about the steps toward near-human intelligence in computers that are being taken almost daily. Why not?

There's been no lack of focus on the scary prospect of smart rodents. In 2000, Stanford's Irving Weissman proposed transplanting human neural cells into fetal mice. The purpose was to study human neurons in vivo, but the proposal quickly turned into a flash point for debate about the ethical issues involved in human/nonhuman chimera. Senator Brownback introduced legislation to criminalize attempts to make such creatures, called the Human Chimera Prohibition Act, though most of its definitions seemed to deal with hybrids. President George W. Bush called for a ban on "human–animal hybrids" in his 2006 State of the Union address. Yet a surprising intelligence enhancement event is far more likely to take place in the AI realm than in animal studies. Although both are far-fetched, the very criteria for AI intelligence enhancement are still very much at issue, whereas physical limitations such as the size of a rodent skull appear to be decisive limiting factors on the effects of any humanized implant.

Anticipatory governance has been defined as a system of governing that is made up of processes and institutions that rely on foresight and predictions to decrease risk and develop efficient methods to address events in their early conception, or to prevent them altogether. It addresses the Collingridge dilemma, a procedural paradox with regard to the social control of new and emerging technologies, which states that during the early phases of development, it is difficult to predict the impact of a technology, but after it has been implemented, the technology is at best difficult to manage.

Anticipatory governance is a common strategy of regulatory frameworks in the basic life sciences. In the extreme, moratoria have been adopted for recombinant DNA technology and for "gain of function" vaccine studies, in each case accompanied by reviews of safety measures and risks and benefits. Sometimes a moratorium is recommended by prominent figures and, though not formally implemented, the proposal leads to systematic reviews, as in the case of gene editing. In a less extreme example, the National Academy of Sciences promulgated standards for the conduct of human embryonic and induced pluripotent stem cell research. One concern that the stem cell research guidelines addressed was the inadvertent "humanization" of nonhuman animal intelligence. Similar concerns about human-to-nonhuman animal neural tissue transplants are being addressed in the context of research involving cerebral organoids.

A parallel conversation about intelligence enhancement is taking place in the world of artificial intelligence. Near-term concerns include the prospect that AI-based systems will upset labor markets that employ large numbers of low-skilled or moderately skilled individuals, such as trucking. Ride-sharing apps have already riled taxi services worldwide. GPS technology is a critical but preliminary foundation for self-driving vehicles. These applications are already largely on the shelf (and in some respects are included in vehicles on the market), and a variety of policy proposals are under consideration, even though none seems likely to prevent the ultimate outcome.

A longer-term worry about enhanced AI is that the inadvertent creation of a radically more intelligent system would be an existential risk to human beings. Various high-profile individuals have expressed this worry, including Hawking, Kurzweil, and Musk. Bostrom has developed the most systematic scenario for such an event, detailed in his book *Superintelligence.* Even short of the doomsday scenario, a partially conscious or fully self-conscious AI would be a world-changing event (even in the absence of clear definitions or criteria for such an entity). Say, for example, that the burgeoning field of machine learning in which algorithms improve with experience is able to replicate a network of neurons, such that the neurons change one another while the system is operating. Those artificial neural networks don't outwardly resemble biological neural networks because they wouldn't be electrochemical, but so far there doesn't seem to be a reason that they couldn't be functionally equivalent, or even better at their tasks. And if they are, then we are led back to the underlying philosophical question about their ability to be self-aware. As the physicist Mark Tegmark observes, "Evolution probably didn't make our biological neurons so complicated because it was necessary, but because it was more efficient—and because evolution, as opposed to human engineers, doesn't reward designs that are simple and easy to understand."[16]

Curiously, however, despite the ongoing anxieties expressed in many quarters about progress in AI, and despite the fact that some form of regulation is often advocated in the abstract, groups formed to address the policy implications (such as Stanford's AI100 group) have not undertaken anticipatory governance efforts with regard to the more extreme outcomes foreseen by Hawking and colleagues. Rather, they are focused on the far more likely disruptions, like disruptions to markets, even if AI is not conscious in any meaningful sense. There is therefore a gap between the most extreme concerns about intelligent AI and anticipatory governance strategies. A project that sought to correct this regulatory gap would theorize that the kinds of anticipatory governance considerations raised on the in vivo side might be useful for the development of regulatory frameworks on the in silico side of intelligence enhancement. It would include a series of workshops with participants who are influential in life science ethics and law, as well as those in the fields relevant to AI. The goal would be to determine whether and how experience with anticipatory governance in the basic life sciences can inform developments in AI that could lead to a threatening form of in silico enhanced intelligence. Papers would be commissioned and a consensus report and recommendations would be prepared. Considering the current absence of any such project, and the immense public interest in these issues, this would be a high-impact effort.

STEPPING BACK

From the beginning, we've emphasized that attempts to come up with a grand theory of the brain have run aground repeatedly, most recently in the mid-twentieth century. The last thirty years or so have focused on experimental interventions, some in the operating room and, increasingly, in the lab. The advent of diverse neuroimaging systems has had a lot to do with this technological turn. More recent contributions have come from fields such as genomics and computing, as well as from improvements in the algorithms, which have gotten "smarter" in light of experimental results, among many other advances. There are few fields of human inquiry that are more multidisciplinary than modern brain science.

But knowing how to be excited without going too far is not an easy epistemic chore. The motivational allure, so enticing for many of us, is a powerful draw. We run on excitement, and the prospect of neuroscientific advances is quite exciting. After all, much of science is drudgery—or being shown to be wrong through failure at the experimental table, in prediction or in test, or in struggling with theory. But the momentum is on the side of discovery, not only because of the scale of research going on but also due to the amount of corporate investment in neuroscience and problem-solving systems. Clear indications of scientific progress include the remarkable new ways to depict the brain, to encapsulate information, and to understand the genetics of neural function and behavioral adaptation. No doubt there's a lot of "neuro hype," but the residue is nonetheless impressive.

Yet we have to caution about the potential for abuse. When we think of the use of neuroscience, we are on the progressive side of the Enlightenment. On the positive side of progressivism, symbolized by the advances in neuroscience, knowledge gained in neuroscientific research is used to advance our capabilities. Diminished pathology and enhanced function are key ingredients of neuroscientific discovery. The darker side is the potential for knowledge abuse, for the use of science that, in the long run, is not in in the interest of human flourishing.

One thing we have learned is that experimenting on the brain not only teaches us how the brain does what it does but also helps us to understand how and why we experience what we do. That observation, confirmed time and time again, is partly why we think of ourselves as philosophical naturalists. So when it comes to what some philosophers call the "hard problem" of how consciousness can arise from a material object like the brain, no matter how complex and marvelous that may be, we are unimpressed. To us, that "problem" is not hard and is not a problem, unless you start out by supposing that reality is so chopped up and radically disconnected that it includes different kinds of "stuff," like matter and mind. But then the deck is loaded, so that a feeling of awe and mystery must lead to something like the so-called hard problem.

What an experimental philosophy leads us to, rather, is a sense of the continuity of reality that is flexible enough to allow different philosophies to be plausible in different circumstances. For example, monism (the idea that there is only one kind of stuff in the universe) works well when I get a shot of the brain chemical oxytocin that can, under suitable conditions, instantly make me feel warm and fuzzy toward my partner (among other things). But dualism is at work when my psychoanalyst and I are trying to remember what it was that attracted me to my partner in the first place.

It's hard to graduate from philosophical concepts that were once powerful but have outlived their ability to adapt to new conditions, including the information provided by experiments in neuroscience. If only we could just let the facts speak for themselves. But we have to admit that facts are embedded in larger conceptual schemes, so instead of a hopeful but naive attempt to "bracket" the baggage of old theories, we say *Full speed ahead*, without being too preoccupied with theorizing the brain. Whatever the prospects for brain simulation, the brain-building we've described plays a crucial role in the strategy of modern neuroscience, which is to let the facts fall where they may and damn the theory. At least for a while.

3

EVOLVING

Our conception of what it is to be human must be tied to an understanding of our evolutionary heritage. The great geneticist Theodosius Dobzhansky said that first scientists, and later modern society in general, see "all things in light of evolution."[1] While that might be an exaggeration, there's no doubt that, whether you love him or hate him, Darwin's influence on the way we see the world and our place in it has been profound. Like other great ideas, however, even our conception of evolution continues to evolve. The one-dimensional trajectory that Darwin envisioned has long since been replaced by the acknowledgment that evolutionary progress is far from linear.

Life on this planet began about 4.5 billion years ago. By comparison, the period between 100,000 BCE, when the first tool-making modern humans emerged, and 10,000 BCE, when agriculture began, is not even a rounding error in geologic time. Unlike these relatively recent changes in human capability, though, the chemical messenger systems that allow our brains to function date back millions and millions of years and are found in both invertebrates and vertebrates. Some neuropeptide systems are found in protozoa, bacteria, fungi, and plants. For instance, insulin, an information molecule, is found in protozoa, bacteria, fungi, and invertebrates. Steroids occur not just in vertebrates but also in plants and invertebrates.

The great twentieth-century synthesis in science, as Ernst Mayr put it, tied Gregor Mendel with Darwin and tied genetics to considerations about adaptation and the larger environment. We are still living with the results of this fusion. However far we have come, biology is often surprising. None of our theories and scientific syntheses could have perfectly predicted a creature like the platypus, and certainly not the human brain itself.

EVOLUTIONARY CONSIDERATIONS OF OUR ORIGINS

Around 4.5 million to 6 million years ago, humans diverged from the proto-chimpanzee and other related apes. We diverged from gorillas some 7 million years in the past and from orang-utans around 12 million years ago. But these early hominins were not a single species; rather, there appear to have been diverse forms of primates on the human tree, around 4 million or 5 million years ago, that probably were closer to the chimpanzee than to us in many respects. Despite how different from modern humans such animals looked, however, at the level of DNA sequences, very few actual changes make up all the differences. Only 1.5 percent divergence exists between us and chimpanzees (although current theory holds that it's not just the actual DNA sequencing that matters but how genes are expressed or not expressed).

Core structures in the evolution of *Homo sapiens* include cranial and dental features and erect posture. These features probably led to greater sharing of food, division of labor, family relationships, and a longer weaning period. By 1.5 million years ago, humans had developed unique and complex behaviors such as seasonal housing and many other forms of shelter, the use of fire, and progressively greater tool use. Bipedalism, or upright walking, was essential for human evolution. *Ardipithecus rami-dus* appeared some 4 million years ago, and skeletal remains of this hominin show a proper heel and a head held upright on the spine, indicating full bipedalism. By the time *Homo habilis* appeared, some 2 million years ago, significant changes in pelvic morphology and further neural expansion continued to control our upright walking skills. *Homo erectus*, with narrow hips, a proper waist, and the ability not only to walk upright but also to run for long distances, spread around the globe.

Upright walking affected behavior in many ways; among other things, it enabled the independent use of our hands. A flexible, opposable thumb made fine hand movements and delicate grasping skills an option, and the development of diverse tools increased our range of motion, action, appraisal, and communicative capabilities. The range of gestures must also have expanded in a context where syntactical structure was first appearing. A wide range of premotor and motor function regions are tied to forming, discerning, and comprehending gestures.

Additional communication tools came with the important development of the larynx. This expansion of the larynx is one thing that distinguishes us from close relatives such as the chimpanzee, and no doubt from some of the hominid species that went extinct. The range of vocal expression expanded with the selective advantage of the larynx. Neck length increased and a tongue reaching into the pharynx allowed articulated speech to evolve. These were all the physical elements needed for eventual communicative competence.

Greater motor control was aided by corticospinal and cortical brain stem neural connectivity, along with developments in key regions of the frontal cortex, such as Broca's area, for the cognitive/motor control of syntax and the expression of propositions. Human language formation, syntax, an expanded larynx, and the right neural circuitry (mediated by regions of the neocortex and a region under the cortex called the striatum that are important for language production) are critical features of humans as they appear today. As the evolutionary anthropologist Daniel Lieberman so nicely puts it, "The larynx is a source of acoustic energy, not unlike the reed in a wind instrument."[2] An expansion of the vocal cords and other motor cortex involvement only facilitated our range of social contact. Our evolution is closely tied to greater social contact, so the development of articulatory skills and voice production vastly improved our communicative capabilities and hence our ability to socialize.

An expanding cortex reflects an evolutionary adaptation in our species, but one should note that the expansion of a piece of other brain regions in other species reflects *their* evolutionary context. For instance, birds—and social birds, in particular—have expanded subcortical tissue, sometimes referred to as the basal ganglia. This region may serve some of the intellectual and social function in birds that an expanding cortex does for mammalian tissue.

Nonetheless, at the level of the brain stem and the cranial nerves (all twelve of them) and of the lower brain stem, structure is remarkably the same across mammals, and indeed across all vertebrates. The big change is in the development of the forebrain in primates. The frontal cortex is almost a third of our brain mass. Through evolution, the visual cortex has expanded, as has all cortical tissue, except the olfactory cortex. Prefrontal and motor cortices make up the bulk of the cortex.

One core feature of the motor cortex is the organization of action. Regions of the motor and premotor cortex are tied to the computational assessment that underlies the approach to and avoidance of objects. The motor cortex sends direct projections to the spinal cord. Our basic behavioral coordination depends on muscle-directed action from the more ventral region of the premotor cortex, along with the functions of the cerebellum. The cortex is rich in cognitive resources and connectivity to both cortical and subcortical regions of the brain. But the large cortex shouldn't be too proud of itself. Invertebrates like the fly have no cortex and yet perform amazing behavioral displays. Social insects can perform dances that indicate where food resources might be found. They have key sensory organs for taste and olfaction, with a small but sophisticated brain that orchestrates behavior and utilizes resources. Insect brains boast a variety of information molecules (steroids, peptides, and neurotransmitters). Sensory systems reflect the diverse forms of adaptation. For us, it is primarily vision; for dolphins, it is audition; for rodents, it is smell.

At one time, the thalamus and cerebellum were seen as mere way stations from the brain stem to the cortex and the cerebellum was seen as simply managing motor functions. These regions are now understood to be richly computational at each level and in each region of the brain. The evolved classical spinal cortical motor pathways are works of anatomical majesty. Long and direct, the organization of the motor, visceral, and sensory systems are multisynaptic. They are at the heart of the organization of action and the regulation of bodily viability, and the bulk of the basic evolutionary plan has remained fairly intact over the past fifty to one hundred million years, across vertebrates.

One more thing about brain size: In a result that surprised many, Suzana Herculano-Houzel showed that there are approximately 86 billion neurons in the human brain, not the 100 billion that is commonly cited. She accomplished this feat of counting through arduous methods, by chopping up human brains (and those of other species, too) and exposing them to a dye to which neurons specifically respond by turning red. After years of work, she reached the somewhat surprising conclusion that human brains are about the same size as those of primates, relative to the size of the rest of the body.[3] Previously, it had been thought that human brains were far larger, relative to body size. Instead, the human advantage is that the cerebral cortex is far larger. Elephants have three times as many total neurons; they just have them in a different place. There is a neat evolutionary explanation for our densely packed cerebral cortex: we figured out how to cook, thereby extracting a lot more calories from our food, which were needed to power the construction of a high-energy brain over many generations.

THE VISION THING

Our species is visual. Dolphins, on the other hand, are auditory, and dogs, our coevolving companions, are more olfactory. If we were just measuring neocortical folds and weight, the dolphin's neocortex would be larger than ours, even controlling for size and body weight. The twentieth-century neuroscientist and psychiatrist John Lilly was fascinated by the sheer size of the dolphin brain, and on that basis he tried to communicate with them, to little avail.[4]

But Lilly was not entirely wrong. Controlling for body weight, the dolphin does have a larger cortical brain mass than our own, and this correlates quite well with the range and complexity of an organism's social relationships. Accordingly, an enormous proportion of a dog's modest brain is devoted to smelling. If one looks at insectivores, the portion of the cortex devoted to olfaction is about 70 percent. But in primates, the visual cortex dominates

TABLE 3.1

HUMAN BRAIN COMPARED TO MOUSE BRAIN

Category	Human Brain	Mouse Brain
Mass of brain	1,500 grams	0.5 grams
Number of brain neurons	86 billion	70 million
Number of cortical neurons	16 billion	14 million
Fraction of cortex that is visual	20%	Almost 10%
Cortical regions involved in vision	30	10
Number of neurons in visual cortex areas	5 billion	1 million to 2 million
Axons in optic nerve	1 million	45,000

Source: Adapted from Koch and Reid (2012).

and takes up a large percentage of the cortex, and indeed of our entire brain. At least thirty regions of the neocortex are involved in vision, representing 20 percent of the cortex. Our brain weighs about 1,500 grams, with 86 billon neurons, 5 billon of which are located in the visual cortex (see table 3.1).

In evolutionary terms, that's a major investment. Genetically, while there are differences across the visual cortex in different species (such as in a mouse compared to a macaque or a human), the design pattern of neural gene expression is remarkably predictable across mammals. This has been the consistent observation from researchers at the Allen Institute for Brain Science.[5] Moreover, the degree of visual cortical expansion and binocularity of the visual systems (looking straight ahead) are core features of our neural or species-specific evolution.

Modern molecular fingerprinting has revealed our evolutionary trajectory in greater detail than was possible at the end of the twentieth century. Ancient DNA is now routinely extracted from extinct species, putting us in a position to effectively reconstruct the genomes of long-vanished creatures. It is not so far-fetched to imagine even growing tissue at some point, as depicted in movies like *Jurassic Park*, if we can find the right frozen mix of preserved material. Analysis has already revealed specific points of divergence in our evolution and distinct lines of related species. Molecular fingerprinting, where it can be used, has therefore added an important element to our understanding of ourselves.

Among others, Svante Paabo at the Max Planck Institute for Evolutionary Anthropology has highlighted the remarkable tools of modern genetic analysis, reconstructing our molecular fingerprints from the past.[6] Our genomic makeup contains about 3 billion nucleotides, and our species is a reflection of the mitochondrial DNA that is passed down in egg cells. There is a high mutation rate through mitochondria, which makes it relatively easy to track human migration across the globe. In a word, we are descended from migrants. All of our ancestors were migratory, eventually roaming over most of Asia and Europe after moving out of Southern and Eastern Africa.

Some five hundred thousand years ago, *Homo erectus* evolved into *Homo heidelbergensis*, a forerunner of both us and the Neanderthals. DNA studies and dating of artifacts tell us that, around two hundred thousand years ago, genetically and anatomically modern humans populated Southern Africa and began to enter Southwest Asia. About eighty thousand years ago, climatic and other environmental changes led to economic development, tool development, and greater social sophistication. Sixty thousand years ago, migration across Africa and Eurasia increased. Migratory travel to North America began about twenty thousand years ago, although there is considerable controversy about the exact age of some North American sites.

The genomic sequences of all living humans are 99 percent identical, so in that sense we are all closely related. Parts of blocks of genomes are fairly straightforward and have many different capabilities for genomic tagging, giving us a sense of both divergence and commonality across all known humans, great apes, and our ancestors—all hominoids. A draft genomic sequence of the remains of three Neanderthals has revealed 4 billon nucleotides, the building blocks of DNA and RNA. Some differences with modern human genomes may have given them a somewhat different metabolism, skeleton, and cognitive capability. The exciting part is the mapping of the DNA sequences and the use of mitochondrial DNA when comparing hominoids. Bones, tools, cranial features, and now molecular fingerprinting all are linked in the consideration of evolution generally, and in the evolution of the brain in particular.

Piece by piece, details of the reconstruction of more than 6 million years of our relatives give a sense of their coming into being. Such reconstruction is set in a body of knowledge that starts with Lucy, an *Australopithecus afarensis* skeleton; nearly 40 percent of Lucy's bones survived to give us our first really startling glimpse of an early hominin. Mary and Richard Leakey characterized *A. afarensis* as a tool-making brain with an ancient hominoid body. We are not the only animals to use tools—crows, for instance, are effective tool users—but no species out there is remotely like us in breadth and depth of inventiveness in tool making. The invention of usable fire was a huge step in our toolmaking development and might date back as far as eight hundred thousand years ago.

We modern humans in our current form weren't an inevitable product of evolution. Modern variants of hominoids competed effectively for environmental space at the same time as us, some thirty thousand to fifty thousand years ago. Neanderthals coexisted and probably interbred with *Homo sapiens*, according to telltale DNA in many modern humans—up to 2 percent. Indeed, recent evidence suggests that the species interbred, in some regions, for thousands of years.

One gene common to both us and Neanderthals is FOXP2, which underlies speech and language and the cognitive/motor regions of the brain (such as the basal ganglia) that are central to forms of learning and behavioral adaptation. In the 1940s, Barbara McClintock discovered that genes are not necessarily fixed in their location on the DNA chromosome.[7] Her work on "jumping genes" was done on the maize genome. She helped to introduce the concept of activator or dissociator gene regulation. She found that genes are altered by events. They can be silenced or activated, depending on what is happening around the organism. DNA, in other words, is under regulatory control by the expression of RNA sequences.

Although Neanderthals went extinct around twenty-eight thousand years ago, they actually had a longer run than we have had so far. The reasons for their disappearance are hotly debated. What we do know is that Neanderthals, in the end, were less successful and that our own version of humans eventually took over their ecological niche. It would not necessarily have taken much time for Neanderthals to be reduced to small pockets of isolated individuals who could not breed their way out of trouble. Their disappearance is poignant when you consider that they were surely smart enough to know that their bands were diminishing. It is a lesson that we might well take to heart, especially as our climate changes.

Neanderthals may have expired as a species, but in addition to their genes, they have left a lot to be remembered by, including burial sites, decorative elements, camps, and tools. Evolutionary biologist Steven Mithen has speculated that the propositional language expression so typical of *Homo sapiens*, composed of sentences logically related to one another, evolved out of song, and that Neanderthals were in fact singers rather than speakers.[8] Reinforcing this idea is the fact that they did in fact leave behind several types of flutelike musical instruments.

This is one of the many ways in which our brains and those of Neanderthals are likely so similar. Musical expectations recruit several regions of the brain, including Broca's area and parts of the basal ganglia, both tied to syntax and statistical inferences. A little bit of disruption goes a long way in listening to and enjoying music. These expectations about what is to come when listening to music arouse further interest in the allure of musical expression. The titillation of an expected verse and note is part of the allure and excitement of musical enjoyment. Indeed, more generally, brain regions that underlie reward expectations, such as the nucleus accumbens in the basal ganglia, are activated by music.

THE BRAIN IN YOUR STOMACH

Peptides are short chains of amino acids. They are smaller than proteins and bind to them. These peptides date back millions of years, and they underlie our standing erect and hence our ability to migrate, due to our walking and running skills, which mark hominoid evolution. Standing erect, the freeing of the hands, and opposable, dexterous thumbs also led to an expansion of our capabilities.

Many peptides in the brain date back 1.5 billion years as molecules that retain and transmit information. One already mentioned, corticotropin-releasing hormone, is best known for its role in the response to change or to adversity. CRH is found in flies, where it regulates circadian rhythms. CRH in mammals is also found in skin and skin pigmentation, in all inflammatory processes, and in the placenta. Information molecules and their migration and use across different biological tissues and different species is only one example of the expanded use of a single piece of biological programming. In this case, a peptide is found and used in many different organisms and tissues, including that of the brain.

Though we mainly see the brain as an organ of thought and consciousness, it is also a regulatory organism. In a constantly changing world, the brain is always on alert. Regulation of information molecules in the brain is less about staying the same (homeostasis) than it is about adapting to or predicting unexpected events (allostasis), the physiology of adapting to change. Yet another example of the brain in context, these processes of continuously adapting to change and solidifying or codifying what is familiar, stable, and expected are critical features in the evolution of the brain and behavioral adaptation. Since we are social animals, the regulation takes place within a social/ecological milieu. The term for this adaptive function is predictive homeostasis or allostasis. In order to give it a wide range of capability, an evolving brain must be predictive and not just reactive, involving both the central and peripheral nervous systems in an endless loop of coordination with peripheral end organs.

The regulation of information molecules is a common theme of our brain's chemistry, entailing restraint and restoration of information molecules and further expression of molecules that underlie behavioral adaptation. In addition to CRH, there is a great range of information molecules, including neurotransmitters, neuropeptides, peptides, and steroids. They range from the commonplace, such as cortisol or insulin, to the less familiar, such as leptin and ghrelin. The brain is represented in every end organ system through the expression of these information molecules. Receptor sites for these molecules are found in the brain itself and in peripheral tissue such as the stomach, heart, skin, and kidneys.

When we talk about the regulatory brain, the word *information* needs to be taken in a somewhat different sense than we usually use it, just as we have become accustomed to thinking of a series of ones and zeros as conveying information in the cyber world. In functional terms, information can be taken as a form of instruction. Because of information molecules, our digestive system is able to consume, digest, and make use of a wide range of food resources. In hominoid evolution, this capacity was coupled with an almost threefold development of the body size. Concurrently, information molecules (such as peptides in the gastrointestinal tract, neuropeptides, insulin, leptin, ghrelin, and growth factor) are critical in fat metabolism, caloric utilization, and protein absorption. Our evolution is closely coupled with transformations in the gut. Long neurons from brain regions like the amygdala project to regions that regulate the gastrointestinal tract.

There are also gut peptides in the brain, where they are called neuropeptides, and they can traverse the whole of what is called the digestive system. Information molecules that promote growth and absorption also shield tissue from degradation over a lifetime of neuronal (and other tissue) activation. In the 1970s and 1980s, scientists found not only that the peripheral system was lined with gut peptides (such as cholecystokinin, insulin, leptin, and neuropeptide Y) but also that their expression in the brain assisted in the organization of appetite and satiety.[9]

No doubt, increased brain size is also tied to rudimentary gut development, and the degree of activity of information molecules in the gastrointestinal tract has led investigators to refer to the gut as a region of the brain, or as a small brain in itself. Peptides are a practical feature in our culinary tastes, a key adaptation tied to neural development and alternate forms of food resources. Oxytocin and vasopressin are two of the better-known peptides. There also are many endorphins and some hypothalamic releasing factors that underlie gonadal steroid and metabolic regulation.

One function of this gut neuronal axis is the experience that we have, in problem solving, of getting the gist of something. Consider the commonplace term "gut feeling." Ralph Adolphs and his colleagues have noted the important role of brain regions such as the amygdala for remembering critical features of events and for problem solving that reflects gut feelings— the gist of the event—because regions of the amygdala or cortical regions project directly to the alimentary canal or digestive tract.[10] That is not to say that the memory resides in the amygdala, just that the amygdala appears critical for attending to what counts (not only what a face might mean but also when to pay attention to it, as gut and circumstance might dictate). This is the sort of problem solving that is good enough in everyday events, rough-and-ready heuristics that guide adaptation. Gut and neural gut peptides might underlie this important adaptation. Thus, one way in which the gut and brain inform each other is via chemical messengers systems such as information molecules from the gut, the same chemical systems that are produced in the brain.

INFORMATION MOLECULES
AND BRAIN CHANGES

We all remember certain big events in our lives. One way this happens is through the steroid cortisol, secreted by the adrenal gland. Cortisol facilitates the expression of the neurotransmitter norepinephrine, which enhances attention. One result of such enhanced attention to external events is the movement from short- to long-term memory. This occurs partly by the lateral amygdala broadcasting the possible significance of the event through a number of regions in the brain. One result is learning and the other is memory. The consolidation of an event into memory involves many regions, including the hippocampus and the neocortex, enabled by neurotransmitters.

Information molecules in the brain underlie the appetitive/consummatory experiences associated with diverse behaviors regulated in the brain. Information molecules like vasopressin are associated with aggressive behaviors and affiliative behaviors; central infusions of vasopressin can increase aggressive social behaviors. Neuropeptides are part of the brain's social signaling systems. When male mice participate in parental behaviors, oxytocin is elevated in anticipation of females delivering pups. Background environments set the condition for neural proliferation in regions of the brain such as the hippocampus, which is critical for memory and, particularly, for the expansion of short-term memory.

The expression of aggression due to the activation of these neuropeptides or steroid signaling systems varies a great deal. Testosterone does not equal aggression, and there are many forms of aggression. A bear protecting her cubs is one example. But testosterone does potentiate muscle mass. Ask your friendly chemical trainer whether it will make you a Barry Bonds.

Regardless of her answer, you should know that it could also make you psychotic. One set of hormones is not enough for any purpose, but the ratio of a number of hormones helps to set the conditions for a behavior in a suitable context, one in which an individual has had experience.

Bonobos have become famous and fascinating for the ways they use sex to resolve conflict. Domestication may be a feature of bonobo evolution. They more commonly walk upright than do other subhuman primates, and they can look into each other's eyes, something that in common chimpanzees is a threat gesture. Perhaps one difference between the bonobo and the common chimpanzee lies in the degree to which oxytocin contributes to and is expressed in the brain and in behavior. It appears that the bonobo's exaggerated lovemaking behavior reflects self-selection against aggression, mediated by differential informational signaling systems.

Since Darwin, the study of domestication has been important in understanding biological adaptation. Human beings have practiced selection for domestication in animal husbandry for around ten thousand years. Now can we link genomic and archeological evidence for our close companions, the domesticated dog or canine. Dogs' common and disparate linkages indicate that they were domesticated at a number of different times and places, maybe as much as fourteen thousand years ago in Europe and seven thousand years ago in Asia. The brains of domesticated and less domesticated animals are very different, especially in the way information molecules affect organ development.

Curt Richter looked at the adrenal glands of domesticated and wild rats in Baltimore.[11] He found that the adrenal glands of wild rats were twice the size of those in domesticated rats. However, one counterintuitive finding stood out: wild rats that were captured on the streets of Baltimore and then domesticated were less adaptive and were worse at simple regulatory problem solving, such as how to find the sodium needed to manage their metabolism when their adrenal glands were removed and they could not retain sodium. Domestication changed features of the rats' brains. Changes in neural function are lifelong events. We are only beginning to understand something about these changes and what they might mean for human well-being.

EXPLORATION AND DISCOVERY

The brain has an inherent ability to keep track of events that matter. Put in the most basic terms, the brain is able to forage for coherence in the course of problem solving. The human capability to catalog and test is built right into our biology through the utter plasticity of neural tissue, particularly at the level of mental and cognitive function. Problem solving is a core feature in the evolution of brain function; it underlies constructing, planning, thinking about the past, and ultimately, reading and writing.

Following millennia during which these abilities were implicit, the past several hundred years of our cultural evolution has revealed the importance of systematic testing and research in comparing various approaches to solving problems. In its most formalized sense, this is what is known as the scientific method. Early investigations into the brain were anchored to anatomy and involved seeing and drawing, dissecting and depicting, and then cataloging the information into an encyclopedia of scientific knowledge about the brain. During the Enlightenment, science was reaching a pinnacle of modern expression, at least so it was thought, and brain science figured prominently, even if it was still rudimentary. Finally, we humans have begun to develop an experimental method for more biological inquiry and an understanding of the brain.

The evolution of problem solving is in part linked to greater access to preadaptive subprograms in the brain that are the backbone for greater and more diverse forms of problem solving, like managing our aversion to certain tastes. Specific preadaptive programs about objects in space, the drawing of causal links between classes of objects (say, between foods and visceral illness), and predictive capabilities about objects over time evolved for narrow but essential requirements (such as acquiring food resources) and were then extended in use. This illustrates that once the subprogram exists, it can be used and reused for many purposes. One mechanism in the evolution of the brain is the increased neural, and hence behavioral, access to specific adaptations for more general use. The view of the brain as a series of sensory motor reflexes, outlined by John Hughlings Jackson and Charles S. Scherrington, has given way to that of an active brain in which the decisional and behavioral expression is constrained by computability in neural tissue and neural design and by its compatibility with the external context.

Many neurotransmitters and neuropeptides underlie the ways in which we forage for coherence: linking events, drawing causal inferences, coping, surviving. Evolution builds on itself. The longer neonatal development period characteristic of humans, relative to all other primates and to most extinct species of *Homo sapiens*, means more time to develop adaptive programs and to build on basic skills. Social interaction and the ability to learn from others also allow for effective use of the building block features of human development. Both learning and natural selection are common themes in genomic evolution. Involved cultural/ecological conditions set the background for neuronal expression, and the brain is the final common denominator for behavior, whether the behavior is domesticated or not. But what dominates is exploration and migration. We are, again, all migrants of different persuasions. The cognitive motor capabilities in neuronal tissue underlie all our migrations.

The anatomical architecture of the motor system embraces the motor neurons and the motor circuitry that underlies behavioral capability. Motor neurons underlie muscle cells, which in turn underlie the autonomic or paracrine regulation and traditional endocrine systems. The somatomotor systems, often called the brain's "go" center, underlie all behavior, from rudimentary reflexes to intentional action. Somatomotor neurons traverse the whole of the brain. Every sensory system is understood in the context of motor neurons, and we even call them sensory-motor systems. J. J. Gibson referred to them as active sensory systems that aid exploration and underlie behavioral adaptation.

The motor systems are at the heart of our activity and are not separated from the sensory systems in everyday action. Both motor and sensory systems are rich in information molecules that underlie exploration. Within sensory systems for exploration through seeing, hearing, and tasting are many different neuropeptides that underlie exploration and discovery of the resources vital for survival, and information neurotransmitters like dopamine that underlie motor systems and cognitive processes. We can move and think, and of course we can think but not move. The brain is an active, metabolically expensive organ, a glucose-eating machine. Much of this energy consumption feeds the central pattern generators, sets of motor neurons. The generators account for the fact that the brain is not passive.

Importantly, neuroscientists do not think of the brain in a rigidly hierarchical fashion, as they once did, with cortex, limbic system, and brain stem, in that order, but it is nonetheless somewhat hierarchical. Behavioral and physiological functions are now understood to be more distributed across the brain. Thus, brain stem functions are embedded in many regions of the spinal cord, are coordinated by midbrain regions of the brain, and are given direction by the limbic system and cortex. The limbic system was once conceived of as a collection of brain organs, including the septum, hippocampus, and amygdala. Taken together, these organs are tied to diverse functions that range from behavioral inhibition to diverse forms of memory and attention.

The older idea of the limbic system was expanded in a conception put forward by Paul MacLean, who proposed a demarcation between neomammalian, paleomammalian, and reptilian to account for the evolution of the brain.[12] This was an interesting perspective. Each level of the neural axis orchestrates behavioral options important to adaptation. The final achievement, in this view, is the expansion of the neomammalian, and eventually the expansion of the neocortex.

As we have transcended the original concept of the limbic system, more and more areas of the brain have come to be included, especially regions that underlie emotional adaptation. Part of what broadened our conception of the limbic brain was the discovery of the chemical architecture of the central nervous system, and particularly the discovery that information molecules in the brain traverse the whole of the central nervous system, including those that underlie emotional or limbic adaptation. Thus, the borders between what is considered limbic and tied to emotional adaptation and what is not have become much more expansive. Limbic areas run the gamut of the nervous system, from first-order sensory information at the level of the brain stem to its projection to the forebrain.

Many of these limbic sites overlap with neuropeptides and neurotransmitters that are engaged in behavioral responses.

These neuropeptides (such as vasopressin, oxytocin, or endorphins) organize responses to novelty and underlie regulatory needs such as thirst and sodium appetite. Indeed, recent results have furthered the idea that this peptide acts to generate thirst and sodium appetite, vital for fluid balance. The circuits that underlie the basic motivational allures include both regions outside the blood–brain barrier (such as the subfornical organ) and diverse regions of the hypothalamus that are often anticipatory in the regulation of the internal milieu, in addition to the amygdala and nucleus accumbens, which are focused on sodium acquisition. Corticosteroids like aldosterone, secreted by the adrenal gland, act to conserve sodium in the brain and to help generate a sodium appetite, all in the service of maintaining fluid balance essential for cardiovascular and cellular health. Moreover, we now have tools for genetic neurochemical activation or silencing of the mineralocorticoid cell-specific activation that generates the appetite for sodium.

Through genetic manipulation of an enzyme, aldosterone can be promoted over competing steroids. Steroids like estrogen, testosterone, progesterone, or vitamin D, for instance, can act on the brain either quickly or slowly. Neuropeptide systems (including CRH or angiotensin, leptin, and neuropeptide Y) are expressed in the central nervous system, and they traverse diverse brain sites throughout the brain. These neuropeptides are part of the central processing rooted in our orientation to basic functions, such as the motivation to ingest or avoid food resources. Modern molecular tools have continued to expand our capability for dissecting circuits in the brain that underlie food ingestion or avoidance or rejection.

For about one hundred years, we have known that core neuronal events are either inhibitory, excitatory, or interneuronal. Antagonist reciprocal interactions across the motor systems reach from the most peripheral sites to the most central sites in the brain—in other words, from the brain stem to the neocortex. Such motor organizations are inherently computational. Motor and cognitive systems are not divorced but are intrinsic to neuronal function. Larry Swanson has elegantly described a blueprint of the brain, in which central pattern generators are both local and narrow for specific brain regions and broad and centrally distributed across broader regions of the brain, depending upon where in the brain they are located.[13] In the brain stem, the central generators may be specific for certain end organ regulation in the gut or the heart. In the forebrain, the central generators are essential in the regulation of behavior and physiology when coping with events in the world. Built into the central generators are computational or cognitive systems that underlie somatosensory or sensorimotor systems in the exploration of the world. Hence, it is not cognition on one side and motor systems on the other. Moreover, it is not just the cortical regions of the motor systems, for example, that are part of the computational carriers, but whole motor systems, from peripheral to more central regions of the brain. In one way or another, every part of the brain is cognitive or computational.

Motivational systems give direction to computation/motor programs, and motivational expression competes within context, circumstance, and capability. Cognitive design and problem solving coevolved in a mosaic of capabilities in a brain focused on making sense of and exploiting the environment. Motor systems, sensory and integrative interneurons, and premotor cortical systems evolved to give direction to behavior in order to organize a framework of coherent problem-solving and goal-oriented behaviors.

In our species, the problem-solving programs themselves are labile and flexible, up to a point. Motor programs that underlie the autonomic nervous system, or neuroendocrine systems, are less flexible, as are basic motor programs for movement at the level of the brain stem and spinal cord. Motor programs across the autonomic systems reach across the whole of the nervous system and underlie physiological regulation (such as hypothalamic pituitary function) and basic biological functioning. The cerebellum, a key motor structure, controls basic motor capability and balance, but this region of the brain is also tied to learning. Learning relationships are crucial to motor control, and expectations and conditioning relate key events to programmatic expectations in the organization of behavior.

With input from all sensory systems moving to the brain in a continuous fashion, one feature of neural design is the reduction of noise. Gating mechanisms funnel and focus attention on what is important and relevant and therefore adaptive. In the real world, again, brains are not on one side and bodies on the other. Despite our ordinary language, in reality there is no separation at all. There are differences in behavioral expressions, behavioral options, cellular capability, and other functions, but they are not dichotomous. Fluidity between brain and body systems is a feature of our evolution and of our success as a species that explores, expands, and has built something we call the culture of knowledge acquisition—within which neuroscience is one form of inquiry among others.

From the arts to the sciences, we rely on our sense of how to get a handle on things, the ways we explore the world. The process is often drudgery, hand over hand, and rarely glamorous. Yet, on we trudge. Underlying our sense of inquiry are the prefrontal cortical neurons, which have bilateral projections to gastrointestinal, spinal, and other systems that organize behavioral responses and underlie our investigations into ourselves, our brains. The blueprint of our brain is connectivity or segregation where necessary. Indeed, some systems are modular, although many are not. The cognitive is not divorced from the contours of action and adaptation.

TOOLS FROM DOPAMINE

Tool use involves continuous expansion of our present universe and its integration with our history. The tools themselves reflect the theories that we impose for understanding our world. Tool making is a feature of our evolution. The degree of cognitive/ motor control and expansion required for tool making is reflected in an evolving cognitive/motor system that traverses the brain. Neurotransmitters, essential for the organization of action and movement, are also fundamental for cognition. Dopamine is one of the basic information molecules in the body and brain that contributes to tool-making capabilities.

Variants of dopamine date back even further than do other molecules, hundreds of millions of years. Dopamine is crucial in the organization of drive and the prediction of rewards. It is involved in the response to incentives and the learning that pervades the organization of action. Dopamine is also essential for the inhibition of action by cortical regions.

Inhibition is a rich concept in the behavioral and neural sciences. It underlies our essential view of neural function. From neuronal firing to gene encoding to neural design to behavioral expression, inhibition plays a role. Loss of inhibition also underlies a number of behavioral vulnerabilities. Mutations in the gene on chromosome 12 that helps to create phenylalanine hydroxylase results in defects in behavioral inhibition and reductions of the tyrosine essential for dopamine production. Adele Diamond and her colleagues showed that children with phenylketonuria syndrome have cognitive impairments in delayed alternation matching tests.[14] These tests are essentially short-term memory tasks, matching what is recalled with what is presented. Cognitive control of behavioral inhibition is consistently found to be tied to regions of the prefrontal cortex and the cingulate cortex. Dopamine serves in both the activation and the inhibition of behavior, in the organization of action.

Dopamine underlies many functions in the brain. For example, interference with regions of the prefrontal or anterior cingulate cortex is associated with changes to the normal response to threat targeted by amygdala activation. Normal memory of fear and appropriate response requires an intact and fully integrated prefrontal cortical region. Stimulation of this cortical region suppresses the amygdala activation in response to perceived or imagined threatening contexts. One result of interference is prolonged excitation, or failure to extinguish the behavioral response after the threatening situation has disappeared.

Dopamine is distributed across diverse regions of the brain, rather than concentrated, because dopamine, like most other transmitters, is in the brain and in the peripheral systems (such as the adrenal gland). Dopaminergic pathways from the substantia nigra that project to forebrain sites such as the putamen and caudate region, sometimes called the basal ganglia, are essential for prosocial motor or habit joint formation. The two dopamine pathways from the brain stem are divided, with one tied to limbic function (in the amygdala, septum, hypothalamus, and regions of the striatum) and another tied to neocortical function (in the prefrontal, cortical, and anterior cingulate regions of the neocortex). Both of these systems are essential for exploration and learning.

The wide array of dopamine expression in the brain, and the many subtypes of dopamine receptors, reveal just how multifaceted this one transmitter is for brain function and behavior.

Dopamine neurons from the midbrain activate specific regions of the amygdala known to influence approach and avoidance behaviors. For instance, dopamine activation (the dopamine D1 receptor) in the medial amygdala is specifically activated for approach or avoidance of a threat.

Dopamine itself is neutral and may be activated under duress or enticement, under conditions of adversity or of provocative pleasure. The range of dopamine involvement is the range of human action, of human expression, performance, and activity. Dopamine is also crucial for sustaining attention. Restoring dopamine levels in Parkinson's patients, for instance, normalizes (at least temporarily) perceptual/cognitive capabilities. Dopamine affects reasoning about statistical inferences in both cortical and subcortical regions (such as the nucleus accumbens) that are critical for motivated behaviors.

Dopamine is essential for any behavioral inhibition or excitation, including expectations of reward. One set of dopamine neurons is activated when expectations and habit formation are not fulfilled. Wolfram Schultz and his colleagues have identified subpopulations of dopamine neurons that underlie reward predictions.[15] Expectations of reward are built into our behavior, as habits. The habit formations that underlie behavior require neuronal systems in the striatum, in which dopamine plays a fundamental role. Statistical computational firing is wired right into cognitive/motor planning, expectations, and disruption. When your plans go haywire, you can blame the dopamine. It underlies exploration, discovery, learning, and inquiry.

Learning is our premium cognitive capability. The continued integration of skills such as writing and reading into frameworks of inquiry reflects our very nature, combining preadaptive capabilities (prepared learning capabilities about objects, of time and place) into more novel and innovative problem solving. Human minds come prepared to group objects into pockets of coherence that are suitable for structure, abstraction, and statistical inference. The utter fluidity of this design is what makes the human brain so flexible and adaptive, so that we are a species that endlessly expands our horizons.

Conditions for detecting familiar and unfamiliar possible danger have not dissipated with the advent of civilization. As a result of the integration of our knowledge of the brain into the broader arena or body politic of knowledge, however, we know how to understand possible dangers better than our prehistoric ancestors did. Damage to the head is one of those dangers, and it is one reason that we have been pondering brains for a long time, starting at least with the ancient Egyptians. Egyptian physicians have left papyri showing the dissection of neural tissue. Later, Greek and then Roman scholars set the conditions for a cultural fascination with understanding the brain. Using our intellect to explore our own brains is one indicator of the essential curiosity of humans.

Our brains are designed for many functions, but social viability is a fundamental one. Many behavioral studies link survival with social competence, which allows us to get a foothold in the world by forging alliances and attachments with others. Not surprisingly, then, all neural behavioral systems are involved in social competence. Amazingly, from a doctrine of the neuron to modern connectome analysis, we have assembled an understanding of the neural function, structure, and design that underlies behavioral adaptation.

Functional circuits still are not easy to discern. But the separation of them in terms of sensorimotor and integrative systems has been further elaborated by computational abilities, present in all neuronal tissue at all levels of the brain. All sensory systems are represented at all levels of the nervous system. The brain is exploratory, and sensorimotor programs support that exploration.

From its peripheral ganglia, the brain is active as opposed to passive, anticipatory as well as reactive. Gene products are modified by gene networks that are affected by circumstance, by learning, by expectations. Diverse gene networks are part of the neural network, visualizing transcriptomes and inferring function across brain regions and functions. It is both the genes and the gene products or genomics that provide modern neural tools for discovery.

4

IMAGING

Suppose that, in order to understand and the brain and to prevent disease or improve the brain's functions, it helps to know what's going on inside the brain. This seems reasonable. But what seems logical isn't always so. For example, how much logical or semantic weight is the idea about being "inside the brain" supposed to bear? If our goal is to know what electrical pathways are activated by a jolt to the skull at a certain three-dimensional location, then we are relying on one sense of "inside." But if our goal is to know how that jolt is experienced, or what behaviors it produces or prevents, another and far more obscure sense of "inside the brain" is invoked.

Not that there's anything wrong with that. Some of the greatest discoveries in science have been built out of a clever experiment. We may find what we're looking for by looking somewhere else, or we may find something entirely unexpected. Or we may not realize that we've been looking for something at all.

TEACHER'S PET

The unique confusions that may be caused by the several ways of expressing what it means to look inside a brain are precisely why we are interested in doing so. As these confusions have vexed philosophers for a long time, it might be thought that modern ways of imaging the brain would help clear things up. In fact, however, they've only added more layers of ambiguity and, at the same time, have implied facts about ourselves that are both disconcerting and require more clarification.

Since 2000, New York University neuroscientist Elizabeth Phelps's lab has been among those conducting a large number of studies showing that people react differently to pictures of faces of their own race than they do to pictures of faces of other races. To put it more precisely, individual neural systems exhibit measurable differences when exposed to different faces, differences that have been identified using functional magnetic resonance imaging or positron emission tomography (PET). This system has been applied to dissecting regions of the brain important for processing facial expression. Meanwhile, MIT's Nancy Kanwisher and many other neuroscientists have used these tools to identify a network of regions that are involved in face recognition and face memory.[1] This essential ability emerges early in development and is loaded with content that is essential for adaptation.[2]

Due to its ability to measure changes in blood flow to various parts of the brain, fMRI has proven to be perhaps the most intriguing of the modern neurotechnology-based brain imaging systems. A research industry has developed around the notion that brain function can be systematically linked to thoughts and actions by observing blood oxygen level–dependent images. Functional MRI scanning takes advantage of the fact that when nerve cells are activated, their impulses metabolize oxygen in the blood that surrounds the cells. The scan records the difference between oxygenated and de-oxygenated blood cells by measuring their magnetic charges, so more active neurons can be distinguished from less active ones. Combine this scanning ability with the assignment of specific tasks or experiences to an experimental subject—the "functional" in fMRI—and the result is a correlation between activated neural systems and mental activity.

PET works by forming a three-dimensional image detected from trace levels of radiation introduced to active molecules. A combination of radioactive tracers and imaging equipment generates 3-D images of organs. Tracers—slightly positively charged particles—are injected either into the bloodstream or into organs. Scanning equipment then records their location to create an image. In addition to detecting problems with the heart or circulatory system, PET has been used to evaluate and diagnose diseases of the brain and nervous system. More recently, PET has been used to identify sections of the brain that correlate with particular cognitive, emotional, and physical activities. Single-photon emission computerized tomography, a less-expensive technology, also uses radioactive tracers to produce three-dimensional images of organs, but the series of cross-sectional slices usually produces an image at a lower resolution than the images produced by PET. Both of these technologies attempt to associate neural activity with specific brain functions.

Face recognition experiments like those done by Kanwisher have identified blood flow to an area called the fusiform gyrus, part of an area named after the nineteenth-century German anatomist Korbinian Brodmann, who stained brain cells. That the fusiform face area within Brodmann's region might be dedicated to facial recognition suggests how important the fusiform gyrus is to our evolution (though the work of neuroscientists like Ed Lein, at the Allen Institute in Seattle, has shown that the area isn't as anatomically distinct as might have been thought).[3]

That specialized areas are devoted to facial recognition suggests how important this capacity is to human interaction; often, this is a matter of life and death. It is not hard to understand that our human ancestors were far more likely to survive if they could pick out a benign face among a number of potentially indifferent or even hostile faces—especially, but not only, in the most dependent and vulnerable period of early childhood. Importantly, levels of the neurotransmitter GABA appear to be reduced in the visual regions devoted to facial processing in the brains of people with autism.

Other imaging technologies that can be used to study the brain include older and more familiar devices that have been ingeniously adapted to new questions. The electroencephalogram (EEG) is useful in identifying the brain's electrical responses to cues. Electrodes attached to the head of a patient record electrical activity in the brain and transmit the data to a computer or a roll of paper, where it is displayed as a series of lines. EEG can pick up changes in the brain, such as seizures, which is useful for detecting epilepsy and sleep disorders, for monitoring patients under anesthesia, and for measuring consciousness. More recently, EEG has been used to study human response to different kinds of external stimuli, as a neuromarketing tool to understand consumer responses to different brands and advertisements, and, when combined with some kind of feedback system, to improve the rate of learning new skills.

Computer tomography scans are a hybrid of computer technologies and X-rays. Frequently used to identify tumors in tissues such as bone or muscle, CT scans also can identify abnormalities in brain structures.

A newer addition to brain imaging techniques, functional near-infrared spectroscopy (fNIRS), detects changes in the optical properties of oxygenated and deoxygenated hemoglobin and other chromatophores (pigment-containing and light-reflecting cells) in the brain. It can be used to measure several types of brain activity in the outer cortex, including motor activity, visual acti-vation, auditory stimulation, and the performance of cognitive tasks. Though fNIRS may be more expensive than other tech-niques and only detects activity in the outer cortex of the brain, it can be preferable in many research settings because it can measure brain activity in more natural conditions. For example, fMRI requires patients to lie down inside of a magnet, whereas fNIRS can be used to measure brain activity while patients are sitting and walking. Intriguingly, fNIRS has been used in decep-tion studies and assessments of deep brain stimulation (DBS) and other therapies, but fMRI is still the big kid on the brain-scanning block.

THIS IS YOUR POLITICS ON fMRI

With all the cool new brain scanning technologies out there, we're entitled to ask how they might pay off in terms of their value for our actual experience. When we think about political figures, for example, what's going on in our brains, and what might that tell us about our attitudes toward them and, most importantly, about their plausible political futures? In Novem-ber 2007, about half a dozen experts in neuroscience, political communications, and applied research published an article in the *New York Times* called "This is Your Brain on Politics." Marco Iacoboni and his colleagues reported on the reactions of twenty people under fMRI, who were shown still photos and videos of candidates and asked to rate them on a ten-point scale from "Very unfavorable" to "Very favorable." Of course, this was almost exactly a year before Barack Obama defeated John McCain for the presidency.

The researchers found that when study participants viewed the words *Democrat*, *Republican*, or *Independent*, they exhibited high levels of activity in the amygdala and the insula, regions associated with anxiety and disgust. This was especially true of male participants when they saw *Republican*, but of course that might have reflected a "liberal bias" on the part of those participants. With regard to specific candidates, reactions to Hillary Clinton were unique in that they showed activity in the anterior cingulate cortex, which the authors said suggested conflicting feelings, "battling unacknowledged impulses to like Mrs. Clinton." They also found that somewhat negative initial attitudes toward Mrs. Clinton among the men became more positive as they watched more videos of her, while women became more neutral. At that time, former New York City mayor Rudolph Giuliani was a very visible candidate on the Republican side. Women liked him better, the more they watched him, suggesting his appeal across the gender gap. Of course, these latter results didn't require neuroimaging. But the fMRI results did suggest that that traditional gender gap (men favoring Republicans, women favoring Democrats) was narrowing, because both male and female participants showed increased activity in the medial orbital prefrontal cortex, associated with positive rewards, regardless of which candidates pictures they viewed. So far, so good.

But when we turn the retrospectoscope on some of the other results, the news isn't good. Showing participants pictures of Mitt Romney back in 2007 led to activity in the amygdala, a sign of anxiety, but that calmed down when they saw him in videos. "Perhaps voters will become more comfortable with Mr. Romney as they see more of him," the writers suggested, though it seems nothing hurt Governor Romney more, when he ran against Barack Obama in 2012, than a video in which he tells the audience he's written off the votes of the 47 percent of Americans "who believe they are victims." Ouch. And Secretary Clinton did not prove more acceptable to male voters in 2016 than she was in 2007. (There's a case to be made that Donald Trump's videos and notorious *Access Hollywood* audio helped him, despite making many voters *un*comfortable.)

There are other ways in which the study turned out to not tell us much about our brains on politics. Looking at pictures of John Edwards, the independent voters in the study had increased activity in neurons associated with empathy, but when the former senator's disloyalty to his wife who was ill with cancer was revealed, several years later, that quickly turned off voters' sympathetic brain cells. As for the two people who ultimately became the candidates in 2008, Barack Obama and John McCain, test participants didn't have much reaction to images of either. Of course, it was the candidates' job to get our attention, which they did at least well enough to be nominated by their parties. And Obama, of course, garnered enough positive attention to get elected. But we didn't need fMRI to tell us that.

Many have pointed out the limits and prospects of brain imaging studies for rendering the mind accessible and understood, and we don't mean to take a cheap retrospective shot at an interesting application of brain imaging. The results that these authors reported were no doubt valid at the time and in the controlled conditions of the neuroscience lab—and those are precisely their limitations. There is a common notion that brain technologies can be taken out of their context and applied to the dynamic embodied and social worlds in which brains operate. That is original sin against the technologies themselves. Had the experiment been repeated daily through the two presidential campaigns, we would undoubtedly have seen the results change. The point is that inferences about future behavior can't be made based on a single dip into brain oxygen patterns. As Sally Satel and Scott Lilienfeld have persuasively argued in their critique of "mindless neuroscience," those behaviors depend on a vast array of variables inside and outside the brain. They are about how we "mind" the world.[4]

DECEPTION

We've depreciated the prospects for technology to improve much on the ancient political brain. If political consultants can't rely on brain imaging to do their jobs for them, what are the chances that brain technologies might be promising for lie detection? Human evolution is rife with cooperative endeavors, as it is with many forms of deception—concealing and misleading others about what we know and what we might do. The evolution of intelligence is rich with an arms race of expectations and tools for expanding our capacities. Over eons of evolutionary time, the same cortical expansion that has enabled tool creation and use has facilitated the arts of deception. Despite frequent references to sex work as the oldest profession, we shouldn't overlook politics as a close competitor.

The implications of reliable deception-detecting systems for the legal system alone would be enormous. After all, unlike political attitudes, which are so dependent on a practically countless array of variables and the vast numbers of ways that we might (literally) mind what's going on out there, lying seems to be a pretty self-contained behavior. Old-fashioned lie detectors are clunky technologies that are mainly useful in intimidating "persons of interest" rather than getting to ground truth. Surely, if brain apps are promising advances in any field at all, they should at least help to advance the cause of truth and justice.

At least two of the technologies we have mentioned that detect brain blood flow and neuronal activity related to lying, fNIRS and fMRI, have been shown to be promising in studies. In theory, the different brain structures that govern truth telling and lying is what could make such systems work for detecting deception. Research done in the early 2000s indicated that truth telling is an automatic reflex governed by structures in the back of the brain, whereas deception involves a more complex array of neurological activity, involving regions of the anterior cingulate cortex and the dorsal and ventral regions of the prefrontal cortex that help both to inhibit the truthful response and to generate the lie. Other experiments suggest that different kinds of lies—in particular, whether a lie is about oneself or about someone else—can activate different neurological events. So if neuroimaging-based deception detection is going to work, the system is going to have to be very specific.[5]

There are other challenges for a lie detection brain app based on fMRI. For example, lying is often closely tied up with recollection. Whether one is lying or not, an accurate answer to *Where were you on the night of January sixteenth?* requires the ability to call up a memory (no PDA-calendar checking allowed). But one study has shown that simple countermeasures can interfere with fMRI memory detection, reducing its accuracy from 100 percent to 33 percent.[6] (Even in a resting state, when the participant isn't thinking about anything in particular, there is MRI activity.) In this study, volunteers participated in a concealed information test in which their neurological responses to both salient and meaningless stimuli were recorded. By simply moving a finger or toe in response to the irrelevant stimuli, the difference between neurological responses to relevant and irrelevant stimuli was diminished. Another complication is that lying in a real-life situation, when there is generally time for preparation, is different from being thrown into a scenario in a lab setting. Neuroscientific lie detection hinges on the concept that deception requires more elaborate neurological processing than the reflexive act of truth telling, but with practice, a complex task such as lying may quickly become second nature and indistinguishable from the behavior to which it is compared. That might be a special problem when measuring the brains of psychopaths who are natural born liars.

Neurological and experiential intricacies of memory also threaten the reliability of neuroscientific deception detection. Memories are not stored like unchanging files on a computer that can be opened at any time. Specific details of memories, and even whole memories, can change and fade, can vary in accuracy according to age, and, if traumatic, may end up suppressed and undetectable. And as Harvard psychologist Daniel Schacter has observed and William James noted as well, there are many types of memory—short-term, long-term, autobiographical, habituation, and others.[7] Different circuits and chemical codes are implicated in all these functions. And it is well known that formal rather than informal questioning (for instance, interrogation as compared to small talk at a party) can alter the very memories it seeks to unearth. Finally, by its nature, fMRI also requires a cooperative subject, since the scans can take from ten to ninety minutes, during which time the subject is required to remain motionless. An uncooperative subject could make it impossible to gather useful data, simply by moving slightly. An action as basic as blinking can cause large fluctuations in fMRI signals.

We hasten to add a philosophical reservation about the whole enterprise of lie detection. Although it might seem obvious that a lie is a lie, sometimes it isn't. Beyond self-deception (a form of false memory), we might quite honestly not be sure when we are lying and when we are caught up in the emotion of the moment.

I love you is often taken to be the ultimate self-disclosing declaration, but add soft lighting, candles, fine wine, a bit of romantic music, and a good shot of the brain hormone oxytocin, and a dispassionate observer might justifiably wonder about the long-term prospects for this relationship.

Certainly, neuroscience has made great strides in identifying brain regions and activity correlated with deception, but several unresolved questions cast doubt on whether neuroscience-based deception detection is ready for real-world applications. Though individual studies have repeatedly demonstrated that truth telling and deception can be differentiated via brain scans, studies regarding the areas of the brain associated with deception have yielded inconsistent results. Also, the same brain regions are, as the biologists say, "recruited" for various functions, so activity observed in areas associated with deception should not be assumed to confirm the presence of deception.

SEEING THINGS

So far, our verdict on brain imaging is laced with reservations. Tamping down expectations about insights into political attitudes based on brain activity isn't much of a challenge. Lie detection proves to be a somewhat more complicated case, but even there we find many promises but not a lot of practical performance, at least so far. For marketers, very specific responses to images like brands and logos might be more promising, yet brain apps for commercial purposes still have to prove their practical worth compared to low-tech measures like focus groups.

Our next example is a far more complex case that not only involves several technologies and complex computations but also challenges our commonsense notions about what it is to see something and to "visualize" it. For discussion, suppose we agree that the invention of moving pictures in the form of film and video has provided us with ways of seeing the world that weren't available to people as little as 125 years ago. That seems straightforward enough. When great painters rendered the human form after meticulous studies of human anatomy, they surely helped us to see the human body in new ways. Similarly, from nature films to love stories, movies help us moderns see the world and our lives in new ways. It is even common for people to describe their recollection of an experienced event as a kind of internal movie, a form of description that would not have been available to our great-great-grandparents.

Now, we might ask whether all of our ancestors up to the late nineteenth century thought about their mental imagery—whether they were seeing something in the outside world or having a dream—as a kind of movie that they could more or less replay in a kind of mental movie theater. Long before the advent of cinema, so recent that it's far less than an eyeblink away in primate evolutionary time, human languages were replete with visual metaphors. Even blind individuals who have never seen color or objects still intelligently and knowingly use language that is rich in visual imagery. The neuroscientist Stephen Kosslyn has found that visual imagery is independent of language and that the visual cortex supports images whether they are being seen or imagined.[8] Linguistic metaphors imply that it is through vision that we gain understanding, as in "Seeing is believing." The history of philosophy might be said to have gotten its start with Platonic metaphors about illumination as intelligence, providing the classical source for the literal enlightenment that has shaped modernity.

These common features of imagistic self-awareness, the terms of our natural language and our sense of civilization, are no accidents. Primate evolution is dominated by the expansion of visual acuity, and this movement toward improved visual attunement influences our physiology, culture, and sensibilities. Kosslyn is among those who have shown that both visual perception and visual imagination correspond to certain common regions of the brain, and the complexity of this system is a reflection of the extent of human visual neural engagement. The expansion of the visual system through its many "experiments" in our evolutionary predecessors is most easily discerned in the part of the brain known as the visual cortex. The brain's capabilities for encoding of visual stimuli are massive, and they need to be. We humans rely in large part on visual cues for just about everything we do, so large proportions of our brains are dedicated to visual processing.

Consider again the example of facial recognition. Because our neural systems are oriented to visual objects and their meaning, a wide array of neural regions is tied to facial expressions. Linking faces and other objects to the wider array of meaning likely occurs through a combination of neural networks. After all, we are endlessly scaffolding against a background landscape in which information is delivered, coded for memory, and embedded in practices. We know that neurons at lower levels in the visual system are sensitive to isolated and specific features in visual scenes. Higher visual areas respond to very specific attributes, but these attributes are remote from the physical stimuli. Instead, they represent increasingly complex concepts, such as the motion of an extended object or the identity of a face. We automatically depict and represent information from the external world within our visual system.

Visual input is so rich in information that, just as we use lots of brain space and neural systems to manage it, neuroscientists can examine the brain's information encoding process to predict future brain function and outward behavior. Visual object identification and recognition is encoded, in part, by the features and functions of the objects. For instance, we know that different objects (for instance, tools versus faces versus animals) are encoded in the brain with regard to the properties and function of the objects (such as employing the frontal motor region for hammers, or perhaps for tool use).

fMRI can help to discern which regions of the brain are active when we are viewing objects, activity that neuroscientists understand as encoding. We can study the relationships between the encoding of such objects and which parts of the brain are activated and then use this information to predict how the brain will behave in novel contexts when it views and experiences new objects, the phase known as decoding. Recent decoding studies have been able to extrapolate which of a series of pictures a participant is viewing based on their neural activation.[9] In other words, when we expect to see an object, that expectation is encoded in our brains, and when we are satisfied that it has actually been seen, that experience is accompanied by a process of decoding that enables recognition. Studying the link between encoding and decoding of events has become somewhat feasible with the use of fMRI, which measures blood flow to the brain rather than neuronal action directly.

Some of the most exciting (and cinematic) recent research attempts to reconstruct the features of a visible object are based on decoding neural imaging. Using mathematical formulae and very fast computers, this research has made it possible not only to discern what a person is seeing from among a select group of pictures but also to identify objects that are unfamiliar to the viewer, or items that are part of the natural world. Using fMRI to reveal unique patterns of activation related to certain events and core semantic categories (like artificial or natural trees and bushes), the aim of these experiments is to specifically reveal what object a person is viewing based on his or her neural response patterns. These decoding studies may eventually be used to extrapolate from the activity of the brain a prediction of what the person is in fact seeing.

THE MOVIE IN YOUR HEAD

But translating images into what they mean to us is not as simple as converting pictures to words. Rather, words like *celery* or *airplane* have different neural mapping patterns in the brain and differentially activate various underlying cognitive systems. The way we apply words to the world reflects categories of function and the kinds of objects. That "semantic" or meaning space is distributed across the brain.

One of the most prominent neuroscientists working in this area is Berkeley's Jack Gallant, who has produced remarkable video reconstructions of visual experiences of participants during fMRI. When we went to visit Gallant in his lab, he explained his goal: because the brain is a mathematical system, he wants to build an accurate model that predicts the real world. He needs to use the same signals that the brain naturally operates on; otherwise, what happens mathematically is that you essentially end up with an answer that doesn't generalize. If you have a really complicated system and you want to predict in that domain, Gallant observed, you have to use the same signals. A sufficiently detailed appreciation of the brain's natural visual signaling system, individualized for each participant based on his or her visual activity while awake, could lead to the creation of a movie that reproduces the visual imaging in dreams, an fMRI-based decoding project that has already been proven in principle in a

Japanese lab.[10] These technological achievements seem to be possible because many of the same brain mechanisms that underlie visual imagery (thinking about an image, whether awake or asleep) are the same as the mechanisms behind visual perception (the images created when we are actually looking at something).

Many important moments of our lives are visual, whether imagined visual or actual visual. Our visual systems are key to our functioning, and they are active even when our eyes are closed, through imagined visual experiences. Given the importance of this system and its effects, it is no wonder that there is a quest to understand the machinery and mechanisms that underlie this experience.

But we need to distinguish decoding visual experience from predicting behavior. It may never be possible to predict which candidate someone will vote for or which beer they will reach for better than a political poll or a focus group could. But there have been large strides in attempts to reconstruct functioning among the visually impaired and to create tools that predict what normally sighted people might be seeing. While fMRI may give us only limited information, and although fMRI findings may at times be overstated, data from these imaging techniques feeds our fundamental understanding of the neural world as it relates to the visual system. Using neurotechnologies in the lab, continued work in this field may yet improve the ability to predict limited forms of behavior.

From a practical standpoint, unless and until neuroscience-based lie detection can reach the near-certain levels of genetic testing, it doesn't have much future as a stand-alone technology. However, a company called No Lie MRI suggests that their product could corroborate witness testimony.[11] Cephos, another company offering fMRI lie detection services, claims that their "lie detection evidence is likely admissible in court" for a variety of reasons, including the admission of brain scans as evidence (as in the legal case *Roper v. Simmons*) and the abundance of previous and ongoing deception-detection research.[12]

As a society, we value truth and think it virtuous to be trustworthy. Neuroscience contributes to the understanding of how trustworthiness might interact with neural judgment and the brain regions or the chemical milieu that underlie these events. Consider those individuals who might be likely to tell the truth: Do they do so because they have a strong will or because they want to avoid conflict or vulnerability? It turns out that it is both. One often hears that the cover-up is worse than the purported transgression. Richard Nixon and the Watergate scandal, Bill Clinton and his sex scandals, Donald Trump's financial transactions, or cover-ups of internet-based political influence campaigns come to mind; telling the truth can be easier. Covering up and continuing to lie requires resources; it is cognitively expansive, consuming, and expensive. The price is obsession with keeping truth from appearing, continued use of neural activation devoted to deception, neural expression, and perhaps neural deterioration. There is no neural free ride.

Consider a set of neuroscience experiments in which trustworthy truth tellers, when compared to liars, experience less conflict and more automatism in telling the truth. Less conflict results in less neural activity as measured in the brain by functional MRI. That may be typical when lying, because it generally takes more cognitive effort to lie—unless of course one is a psychopath or is so convinced by the lie that it serves some greater goal. Such considerations are reflected in what Joshua Greene and Joseph Paxton have called the "grace hypothesis," the idea that measuring honest moral decisions again requires less cognitive control in brain regions like the prefrontal cortex than measuring dishonest moral decisions.[13] Individuals who are willing to lie for gain require more cortical activity. That should not be surprising, as difficulty of decision-making should reflect greater cortical expression and activation. Neuroscientific considerations will not replace our notion of moral character. Rather, they will provide a material understanding of how our brain underlies ideas about the behavior of virtuous people in suitable social context and cultural evolution.

WHODUNIT

Proving that a defendant committed a criminal act requires showing both that the physical act in question took place (actus reus) and that there was intention to commit the act (mens rea). Intentions are mental activities, so mens rea provides a nice testing ground for the practical application of deception-detection technologies. In the law, mens rea typically involves a gradient of four degrees of intention: purpose, knowledge, recklessness, and negligence. Although current neurotechnologies cannot retrospectively reconstruct a defendant's supposed intention at the moment of a crime, it is possible to tell the extent to which a defendant is capable of forming intentions and argue that this ability bears on the charge in question. For example, if a defendant is shown to lack the capacity for intentional action (say, because of a severe brain disorder), his or her lawyers could argue that this inability to form intentions existed at the time of the crime, thus mitigating the charge.

Studies have attempted to isolate the brain regions and patterns of neuronal activation associated with forming intentions, despite the obvious challenges of eliciting and measuring unintentional actions to compare with intended ones. To circumvent this problem, a 2004 study had participants perform the simple action of pressing a button and focusing either on the intention to perform the action or on the action itself. In 2007, a similar study suggested neurological differentiation between intention and execution of an action. The University of Pittsburgh's Peter Strick has demonstrated the connections from the premotor cortex down into the spinal cord.

These studies begin to show that the biological locus of intention is in the presupplementary motor area, among other areas, and that this area makes connections with other regions of the brain, allowing it to influence the formation of intention and the execution of the action. When the presupplementary motor area and associated regions are dysfunctional, complex actions can be performed without intentional awareness. A brain region called the angular gyrus has been correlated with the capacity for reflection on intentions as they form, and a disease or injury causing damage to this area could lead one to be unaware of one's intentions until they manifest as actions.

Traumatic brain injury and the use of medications such as antidepressants also could affect mens rea in a legally significant manner. Attorneys have claimed that drugs like Prozac compromised their clients' mens rea, at which point scientific experts usually weigh in on whether that could be the case. Traumatic brain injury that involves damage to the prefrontal lobe can compromise a patient's ability to incorporate moral reasoning into his or her actions, and it has been hypothesized that if imaging could show this type of damage, it could affect legal decisions. In *State v. Idellfonso-Diaz*, a case involving expert witness testimony about MRI, PET, and EEG evidence, the appellate court held that "expert testimony regarding diminished capacity is relevant only when it can show that a defendant lacked the capacity to form the culpable mental state due to mental disease or defect."[14]

There's some evidence that juries are impressed by biological evidence like a brain scan, but before we get too excited about neurotechnologies in the courtroom, it's important to note that U.S. courts set strict rules on what kinds of "scientific" evidence juries even get to see. Law students learn that *Daubert v. Merrell Dow Pharmaceuticals, Inc.* set the current standard for allowing only scientifically valid and relevant information in expert testimony. Courtroom statements by experts must meet a high bar for what counts as "scientific knowledge": a fact or theory that is well grounded in the methods or procedures of science.

Does brain scanning meet the *Daubert* test? Not according to the vast majority of neuroscientists. Brain scanning studies often have few participants, many drawn from undergraduate student populations, a sample that does not necessarily represent either the population at large or the defendants to whom the studies are intended to apply. It is also unclear how to apply research data to individuals in a courtroom setting, since scientific findings are typically averaged and statistically grouped. Because there can be greater variability between two individuals within a given group than among individuals in different groups, determining, for example, into which statistical group a defendant falls proves challenging. So, getting back to mens rea, fMRI may identify similarly compromised brain activity in two people without being able to specify exactly how much capacity to form intentions one or the other has.

CONSUMING IMAGES

If brain imaging for political forecasting or lie detection doesn't hold much promise, what about neuroscience for advertising? Cutting-edge psychology has a long record in marketing. In the

1920s, Sigmund Freud's nephew Edward Bernays brought psychoanalytic principles to Madison Avenue, and John Watson tried to do the same thing with behaviorism. Will neuromarketing be next? Some advertisers hope that neuroscience can help them figure out how to spend less on attracting consumers and provide new information about getting people to open up their wallets. The traditional methods for gathering information about an individual's preferences—market tests, simulated choices, questionnaires, and focus groups—often yield different results for the same questions. Perhaps brain imaging can both reveal and appeal to peoples' subjective preferences, even if they don't want to disclose them or they aren't conscious of them.

"Willingness to pay" studies determine how much participants would pay for a particular experience, such as eating a particular food, which represents an expression of the experience's expected utility. Willingness to pay has been correlated with the medial orbitofrontal cortex and the prefrontal cortex, which also are associated with anticipation and experience of reward. But we again run into the problem that haunts lie detection: because single brain regions partake in myriad cognitive processes, cognitive processes cannot be reliably deduced from regional brain activity alone.

For example, a study using magnetoencephalography during a simulated shopping experience traced the brain processes associated with product choice.[15] One finding was that participants showed activity in the occipital cortex, a region that is involved with working memory. Then, product choices were compared with memories of buying, using, and seeing advertisements for the product in the past. At 400 milliseconds after the stimulus, male brains showed higher activity in the right temporal lobe, indicating the use of spatial memory, whereas females used the left parieto-occipital lobe more, indicating a choice based on knowledge of the product. If these results bear out, it's not clear how marketers could use them. Advertising campaigns already vary depending on the sex of the target, and there are obviously lots of other nongendered factors that influence purchases.

The scientists also compared predictable choices, where the participants were already familiar with the product and expressed a preference for it. Not surprisingly, those buying decisions were made faster. They involved the right parietal regions and, later, activity in the left prefrontal cortex, possibly indicating comparison of the presented product with memories of experience with the product. Unpredictable choices, where the participant is not familiar with a particular product, were made more slowly. They involved the right inferior frontal cortex, and then the left orbitofrontal cortex, suggesting that saying the product brand name out loud and appreciating its seeming convenience are involved in the decision. These findings suggest that choice depends on complex processes involving numerous brain regions that could be susceptible to influence at many levels.

Once products have been released to the marketplace, studies focus on correlating consumers' brain activity with how useful the product seems and how useful the buyer can be made to feel the product is—its "induced utility." These postmarketing researchers want to understand the neurological relationships among what a consumer expects to get out of a product, what he or she actually gets out of it, and the likelihood that he or she will purchase the product again. But the physical limitations on a participant undergoing fMRI scanning, such as being confined to the tube within the magnet, largely constrain such studies to looking at product pictures and advertisements. That's an unreliable substitute for actually being in a store. Even during online shopping, the physical confinement of the MRI device isn't like lounging at home.

In the 1960s, social psychologists found that the more pres-
tigious degrees a lecturer was said to have, the taller students perceived him to be. Similarly, one emotional factor in purchasing
decisions that does seem to be measurable in neuroscience experiments is the influence of our expectations on our experience of
a product's desirability. A study found that price-based expectations of utility affected medial orbitofrontal cortex responses to
the experienced utility of tasting wine; that is, when participants
believed the wine to be expensive, their brains signaled high levels of experienced utility, and when the same wine was described
as being inexpensive, participants' brains showed less activity in
this region.[16] Similarly, participants have exhibited a greater placebo response to analgesic "medications" they believed to be
expensive. Expectation-inducing cues can be manipulated (and
studied) by adding phrases like "New and improved" to a product or by changing its price. When these things happen, brain
chemicals like dopamine are released in regions of the brain in
a way that corresponds to these perceived utilities. Many experiments have demonstrated the relationship between expectations
and dopamine release.

EEG studies have shown that participants watching televi-
sion advertisements display different brain activity as a function
of whether they remember the advertisement ten days later. The
left frontal areas showed higher activation in those who memo-
rized the commercials, as did the anterior cingulate cortex and
the cingulate motor area, suggesting a role of emotion in mem-
orization. Advertisements and logos also have been studied in
the context of neurological responses to pleasure, and particular
patterns of neuronal activity have been correlated with specific
indications of preference and aversion. When one group of participants was presented with subliminal presentation of brand
logos, they had higher levels of neural activation in the medial

prefrontal cortex. They also selected their "rewards" more impulsively. As we saw, food and political candidates have been studied for neurological indications of pleasure and attractiveness.

These studies suggest that products themselves, as well as marketing tools like commercials and logos, could be designed or even reverse engineered for likability in accordance with brain signals. Many companies now offer neuromarketing services. NeuroFocus, for example, utilizes the Mynd headset, a dry "wireless full-brain EEG measurement" device that is claimed to "revolutioniz[e] mobile in-store market research and media consumption at home."[17] NeuroFocus uses various strategies, prominently including neuroscience, to optimize branding, product development, packaging, the in-store experience, advertising, and entertainment. A company called Neurosense laims to be able to identify the "hidden triggers of consumer purchasing behavior" by focusing on "System 1—nonconscious or fast thinking processes on which most consumer decisions are based."[18]

We can't speak to the validity of these claims, but there's no question that some very intriguing neuroscience is being done in the context of marketing and consumerism. At the University of Pennsylvania, Mike Platt and his team study monkey neurology to determine which brain regions are involved with a decision to stay in one place or to explore another area.[19] The posterior cingulate cortex is active in foraging for a new site, and this perhaps also operates in the brains of shoppers deciding to move to the next display case or of the creative corporate head willing to dive into a new product line. Another Penn group, under Gideon Nave, has shown that a single dose of testosterone can cause males to desire prestige products, perhaps a form of display that other species' males accomplish by means of physical traits such as exotic plumage.[20] Unfortunately, that testosterone boost also was found to cause deterioration in decision-making skills—or maybe that is precisely what consumerism in some measure depends on, especially when beer commercials target young men.

Worries about the loss of privacy due to neuroimaging are important to contemplate but are greatly exaggerated. Our thick skulls are still hard to probe, especially from a distance, and even the most provocative experiments with fMRI require an exceptionally cooperative participant who doesn't suffer from claustrophobia. Worries about privacy are far more appropriately directed at external behaviors that might actually be meaningfully monitored, such as minuscule muscle twitching or online social networking, but those technologies are here already, and they exploit ancient unconscious behaviors. If and when long-range brain scanning can be done, societies will be plunged into far-reaching conversations about consent and—unfortunately more likely—into control by authoritarian regimes for which consent is beside the point.

5

ENGINEERING

Technological manipulations of animal brains to test behavior can have disconcerting results, both for culture and for ethics. In an experiment involving a mouse mating encounter, a male mouse crept up to the female on the other side of their cage, mounted her, and proceeded to copulate. After a second or two, he wandered away. The female followed him, he turned, and she briefly brushed his snout with hers, a typical mating gesture. He circled around and again mounted her. But then the encounter worked out differently. An experimenter activated a fiber-optic cable implanted in the male's brain, streaming a light-sensitive protein called opsin through particular cells, activating them. In an instant, the male viciously attacked the female, jumped on her back, tore at her with his forelegs, and bit her neck. Then, just as quickly, he ceased the attack and returned to his original corner. The laser light running into his brain had been switched off. Experiments to better understand the neuroscience of aggression using new genetic tools have been conducted at a number of laboratories.[1]

OPTOGENETICS

A common target of these aggression experiments is a brain organ about the size of an almond, the hypothalamus. The hypothalamus is responsible for many basic processes in vertebrate animals, including basic circadian rhythms like sleep and appetite. Since the 1940s, when the Swiss neuroscientist Walter Rudolf Hess used electrical stimulation in his experiments (he won a Nobel Prize for his work), it has been known that the hypothalamus is also crucial in mating and, and as demonstrated by the MIT researchers, in highly aggressive behavior. Using a remarkable new lab technology called optogenetics, the ventromedial hypothalamus in the mouse is stimulated by genes and light, and the mouse's behavior looks very much like a sudden, violent rage. What is especially disconcerting about the video is the vivid demonstration that the same brain structures that relate to sex are involved in violence.

Yet we should be careful about jumping to the grim conclusion, based on this sophisticated and highly controlled experiment, that sex is violence, especially in human beings. Biology doesn't allow us to "read" human nature, not in mice and certainly not enough to make inferences about love and war. It turns out that the aggressive behaviors associated with the ventromedial hypothalamus are directed not only at females but also at other males and even, in one experiment, an inflated glove. Nor is the result of the experiment a surprise, however upsetting the results might be to the novice viewer, as the functions of the hypothalamus have been known for a long time. What is new is the ability to target just the right clutch of thousands of cells using optogenetics, and what is surprising is the ease with which that can be done in a well-equipped laboratory. All it takes is the right tool.

We need to avoid being prisoners of our language or our engineering tools. When we think of engineering, we tend to think of modifying a structure. In neuroscience, structural modification is sometimes the right way to think about doing something with the brain, but not usually. The brain is a stable system, but as an organism it is also undergoing constant modification, from both within and without. A familiar flight of fancy in both philosophy (Descartes) and science fiction (*The Matrix*) is that we might, after all, be brains in vats, deluded by some malign force into thinking that we actually live in something like the world we seem to perceive. That brains-in-vats science fiction concept relies on a structural model of the brain, so that all you need to do to get the outputs you want is to directly modify it. This would obviously be a pretty neat trick, especially for brains that have not evolved to have vats as their environment. Suppose, however, for the sake of argument, that we are not brains in vats. In that case, we will have to think about engineering the brain very differently.

EPINEUROMICS

Although many brain scientists think of modifying the brain as a simple structural engineering challenge, we disagree. To make our view clear we have coined a new term, *epineuromics*. The suffix *-ome* has come to be used in molecular biology to refer to a totality, such as the entirety of the human genetic code or genome, including both the coding and noncoding DNA/RNA sequences, the study of which is called genomics. The prefix *epi-* denotes that which is beyond or external. Thus, epigenetics refers to trait variations that are not caused by changes in the genetic code itself. It is now a commonplace that one can hardly study genetics for anything more than highly limited purposes without also studying epigenetics.

Neuromics is the study of the totality of neurons in an organism. It is recognized that the nervous system cannot be fully understood without an account of its interactions with influences that are independent of it. So our suggested term, *epineuromics*, refers to the study of neurons and their interactions, including environmental influences. Epineuromics captures the fact that the brain can be changed within a certain range. To modern brain scientists, this seems obvious, but it took a lot of time for the realization behind epineuromics to become a theme that brings together the many fields of study called neuroscience. Once, the brain was thought to be largely unchanging, but since the mid-twentieth century, countless studies have shown that the brain's structure and function are subject to amazingly subtle modifications. These studies include work on injury, learning, vision, pain, prostheses, computer interactions, and many other subjects. One of our favorite philosophers, the grandfather of modern neuroscience, William James, coined the term *plasticity*, the notion that the nervous system is seen as "weak enough to yield to an influence but strong enough not to yield all at once."[2]

Neuroplasticity is the driving force behind epineuromics. Brain development is irreversibly changed by early life, from visual experience to nutrition to learning environments. Canaries' new songs are accompanied by new neurons, stress leads to cell death, and adult tree shrews develop new cortical cells. Both Prozac and physical activity like running contribute to neurogenesis in hippocampal tissue. As we baby boomers age, we are encouraged to spend more time on treadmills. It is not surprising that running capability is built into the bone formation of our feet. Our great ape cousins are less capable in this regard.

We are slow compared to many other species, but we evolved muscular footgear for a rapid walk and speedy run. And hunting together, an activity in which social contact, cooperation, and cohesion are essential, ripened into an evolved social brain and hands that can be used as tools—as well as running feet, just in case.

Running and other vigorous activity contributes to enhanced learning, shown in a region of the hippocampus called the dentate gyrus, and also results in greater expression of neurotrophic factors that enhance long-term potentiation of neurons. This neurogenesis is tied to the formation of associative strength in memory formation. Brain repair is a continuous property of some neural tissue. Growth and regrowth of tissue is vital for functions like birdsong.

Although we might think it would be best to make our brains coldly rational—if we could do that, like some primitive electronic calculator—that would be a mistake. The emotions have a cognitive function. They strengthen the environmental link in our decision-making. They are a crucial part of what James called the stream of consciousness. Without emotions, we couldn't make our way in the world. We concur with James that, despite long-running debates about the degree to which emotions are cortically or subcortically generated, they are both. Since Aristotle and through Dewey, many philosophers have observed that the emotions are not simple ejaculations of feeling but that they have a cognitive function, organizing our thoughts and actions.

MODULATING THE BRAIN

Modern descendants of the work of Galvani and Volta, electricity-based brain therapies are intended to modulate the ways in which neurons fire and signal each other. *Modulation* is a word from the world of electronics. As we talk about the various ways to modify the brain's electrical signals, we need to keep in mind the idea behind epineuromics, that the brain produces electrical signals but is not simply an electronic device. Its electrical signals can be detected, interpreted, and modulated, but ultimately the brain will have to interact with its environment. So experiments are very different from innovations in the real world. Like brain imaging, which has led to some amazing highly controlled experiments, brain modulation in the world outside the artificial setting of the lab is going to have to take into account the way the brain works in an uncontrolled environment.

Then there are basic practical considerations about the value of investing in systems like this on a large scale. Will the expense of a technology like this be worth the price of implementing it? If it takes forty minutes to learn a new skill the old-fashioned way, will the ability to cut that time in half be cost-effective? Could those funds be spent in a way that ultimately might be more valuable for the organization that is training workers?

Compared to feedback systems, some form of inexpensive external electrical stimulation would be a much more straightforward approach to enhancing brain function—if it's safe and if it works. By targeting the dorsolateral prefrontal cortex, transcranial magnetic stimulation (TMS) has benefited some severely depressed patients, leading people to wonder whether the electromagnetic field it produces might be used not just as therapy but also for cognitive enhancement. Perhaps TMS could bring people with statistically and physiologically "normal" cognitive function to a higher level of intellect, at least for brief periods. Transcranial direct current stimulation (tDCS) also is being studied and marketed (perhaps prematurely) for anxiety, insomnia, and eating disorders. And if a person with a disorder can be brought up to a "normal" level, then perhaps well-functioning people could be made to perform even better. That familiar late-afternoon fogginess that many of us experience might be zapped away, and Sudoku scores could be nudged upward.

Apart from therapeutic possibilities, there's some evidence that TMS can enhance cognition, including the findings of the Defense Advanced Research Projects Agency (DARPA) and its "accelerated learning" studies,[3] but tDCS is different. It works off a nine-volt battery, far less than the current produced by TMS, and it's not clear how that modest zap can get past the skull. There's also a lot of uncertainty about the exact conditions needed to get the desired result, including the right placement of the electrodes with respect to the targeted region. Yet the equipment needed for tDCS is modest—TMS devices cost thousands of dollars but tDCS devices are marketed for a few hundred dollars or less—and the process seems safe over the short term if it's supervised. However, it also can cause brief respiratory paralysis and may induce long-term changes that can't be measured yet.

tDCS enthusiasts claim that its simplicity makes it an inexpensive way not only to treat mental disorders like depression but also to dramatically improve cognition. Do-it-yourself tDCS bloggers, sometimes using homemade devices and sometimes using those advertised for depression, describe improved performance. These DIYers compare notes on apparatuses, targeting, and minor side effects, such as an itchy scalp where the pads are placed. But an analysis of dozens of tDCS studies found no significant enhancing result, and another even found a slight lessening of IQ scores in the period immediately following exposure. So although safe, cheap, and reliable external neuromodulation should be the ultimate everyday neurotechnology, the jury is definitely not in quite yet.

Another noninvasive method of brain stimulation, ultrasound, may offer safety benefits over deep brain stimulation. Ultrasounds are sound waves with frequencies higher than the upper audible limit of human hearing. Ultrasound is no different from normal, audible sound in its physical properties, except that humans cannot hear it. This limit varies from person to person and is approximately 20 kilohertz (20,000 hertz) in healthy young adults. Ultrasound devices operate with frequencies ranging from 20 kilohertz up to several gigahertz. It's used in many different applications, from improving the quality of metals in manufacturing to detecting objects or measuring distances, as in sonar. For many years, physical therapists have used ultrasound therapy to treat bone and tissue ailments. Pulsed ultrasonic technology combines short bursts of ultrasonic energy with electronics that sense a return signal after each burst. An image is built out of those return signals.

So ultrasound is used both to target and modify materials and to get feedback from them that can create visual representations. Your first baby picture might be an ultrasound image of yourself as a fetus in your mother's body. For medical imaging, the spatial resolution of ultrasonic signal processing may surpass that of both TMS and tDCS. It also avoids the complexities of genetic modification inherent in technologies like optogenetics. Can ultrasound also change the brain in ways that can enhance cognition?

In a lab funded in part by DARPA and the U.S. Army, neuroscientists envision the eventual development of in-helmet ultrasound neuromodulation devices. But ultrasound's potential to damage tissue necessitates further research on safety. When warfighters are involved, the balance of risks and benefits may be different from that for other vulnerable groups. Historically, military applications of new technologies have often pushed the bounds of acceptable risk. For example, a commander who believes that his forces are about to be degraded by enemy action may be able to justify the use of a new drug or device if he or she has reason to believe it may be effective and if there are no superior alternatives. By their very nature, combat situations often present a level of existential risk that changes the risk–benefit calculus in ways that do not apply to noncombatants. (We will present more on the brain and what we call "neurosecurity" in chapter 6.)

THE "OPTO-MISTS"

Even if external brain stimulation techniques like TMS, tDCS, or ultrasound turn out to have widespread applications for therapy, and perhaps for enhancement, electrical and sonic devices have a major limitation: they don't seem to be able to stimulate specific types of neurons and leave others alone. Genes routinely do that by making cell-specific proteins. So any technological system that can get brain cells to make certain proteins and then track the proteins as they move around the neural network would allow neuroscientists to associate systems with behaviors, something that has long eluded experimenters. Of course, that level of precision might not be needed for treating depression, or perhaps even for enhancing learning, so TMS, tDCS, or ultrasound might be fine for practical purposes like that. However, they don't help researchers learn about the functions of particular neurons and their pathways at a basic level. That kind of knowledge could also make for a lot more control of behavior.

One biotechnology that has excited both scientific activity and ethical debate is the creation of stem cells, highly potent cells that may be obtained from embryos or from "adult" cells engineered to resemble those found in nature. In theory, they could be used as part of a new kind of cellular therapy. Neural stem cells have been identified and studied in certain structures in the adult brain, like the hippocampus, and have been found able to turn into the brain's major cell types. By their nature, stem cells compensate for the loss of normal cellular functions, so it seems plausible that someday neural stem cells could be introduced into a living brain for therapy, or even for enhancement. Disease targets could include multiple sclerosis and spinal cord injury as well as degenerative diseases like Alzheimer's, Huntington's, or Parkinson's. But putting potent cells into people has its own well-recognized risks, including tumor formation, the possibility of cells migrating to the wrong sites, transplant rejection, and the usual risks of surgery and infection.

Hence the excitement about optogenetics, most often associated with Stanford's Karl Deisseroth. Using proteins called opsins, normally found in the optic nerve, to tag specific neuronal systems and sites, optogenetics allows neuroscientists both to observe neurons and to influence their expression in specific regions of the brain. Optogenetics is a brilliantly simple concept, though it took years to develop so that it would be useful in a wide range of experiments. Neurons in lab animals are made light-sensitive by inserting a gene that produces one of the opsins, molecules that are highly conserved in evolution and are critical to vision in all mammals. The opsins are then turned on and off with a fiber-optic light source in the brain. They not only can be visualized but also can be pushed into one neural pathway or another.

Previously, only one neuron at a time, or at most several, could be observed firing, and the paths they traveled remained largely mysterious. Optogenetics has changed all that. It has allowed experimenters to investigate and control neural systems, with quite remarkable results. In some of the more impressive experiments with rodents, well-fed creatures can be made famished, and rats can be made to exhibit fear responses to situations they've never been in before. Certain behaviors are produced, such as those that count in the mouse as fear, anxiety, hunger, or aggression.

Skeptics might doubt that lab animals like mice and rats can be a model of human experience, but looking at films of optogenetic experiments, it's hard to deny that something important is going on. In one series, a set of neurons that are associated with a certain scary experience is turned off, and then turned on again, with obvious changes in the mouse's exploratory behavior—avoidance of that part of the cage where the frightening incident took place. And that is barely scratching the surface. Optogenetics signifies the convergence of various brain apps made by human beings that take advantage of the clues left by nature's own apps.

Optogenetics lets scientists upgrade and downgrade the activity of neurons in a very precise way so that the activity can be correlated with obvious responses. Fear behavior is a favorite of experimental neuroscience because it's pretty easy to see when it's going on, without getting into legitimate but often secondary philosophical questions about how the mouse "experiences" fear. (Wittgenstein put this "other minds" problem into perspective when he wrote "Just try—in a real case—to doubt someone else's fear or pain.") When certain brain chemicals are more present in the brain, they can be measured. So, for example, when the corticotropin-releasing hormone—a neuropeptide linked to the stress response—is upgraded in the amygdala of a mouse brain, there's more of a trend to be wary, fearful, or socially avoidant. In a set of neurons called the nucleus accumbens, the same chemical can be manipulated so that it generates behaviors associated with appetite. Both can be accomplished with the ingenious combination of genes and light called optogenetics. For instance, silencing CRH pathways from the amygdala to the bed nucleus of the stria terminalis disrupts sustained fear. Optogenetic manipulation of CRH in the amygdala thus can either enhance or degrade threat-related behaviors.

Conditions of adversity facilitate GABA neurotransmission in CRH neurons of the amygdala and bed nucleus. These are important in modulating the expression of CRH in response to fear- or threat-related events. The neurologist Kurt Goldstein suggests that "anxiety has no object," but fear does. Michael Davis considers this distinction essential when dissecting brain regions that are critical for fear and anxiety.[4] Fear is more tied to the amygdala; anxiety to the bed nucleus of the stria terminalis. The lateral amygdala is critical for threat-related behavioral response. Synaptic changes occur in this part of the brain in response to threat, as well as in the central nucleus of the amygdala and the nucleus accumbens. The hypervigilance noted in anxious individuals reflects synaptic changes—changes in the lateral prefrontal cortex underlying working memory and in the medial prefrontal cortex underlying memory extinction.

Regions of the anterior cingulate cortex or the prelimbic region of the prefrontal cortex may be critical for extinguishing fear and anxiety-related behaviors. The preservation of responses to threats that are no longer an environmental feature to worry about can occur when this region is compromised. Ivan Pavlov himself envisioned such a role for cortical tissue, but it was Maria Morgan and Joseph LeDoux who provided evidence for such critical roles in extinguishing memories of fear and anxiety-related events.[5] Regions of the brain critical for extension from fear or anxiety-related events include the prelimbic prefrontal cortex, the hippocampus, which is tied to hypersensitivity, and the amygdala, which is tied to context.

A major question is who is vulnerable to anxiety disorders such as post-traumatic stress disorder. Two people can have the same experiences in trauma-evoking events such as combat, yet one has the disorder and the other does not. What are the vulnerabilities? And just as important, what are the ameliorative factors? We know much more about the ameliorative factors than we do about the vulnerabilities, but we do know that regions like the hippocampus and amygdala are altered differently. The hippocampus can decrease in volume, whereas the amygdala expands under conditions of adversity.

An experiment by Chung and colleagues suggests that the amygdala and the hippocampus are networked in such a way that emotions like anxiety, felt in the present, are strongly related to remembered negative experiences.[6] The researchers obtained consent from epilepsy patients who were being prepared for brain surgery to take continuous recordings of their brains' electrical activity for a couple of weeks. The patients also kept track of their moods. With the help of computer algorithms to crunch the data, the interactions between the fear-mediating amygdala and the memory-assisting hippocampus was strongly associated with low moods in most of the patients.

All of this isn't being done just to enable interesting lab experiments. If anxiety pathways and their related chemicals can be better understood, they could lead to therapies for people who are paralyzed by their disease and for whom current treatments haven't worked. One option is some form of brain stimulation to interrupt the circuit.

THE DOCTORS DREADD

When many of us think about designer drugs, we think of new versions of MDMA (the recreational drug known as ecstasy), but neuroscientists are also designing drugs to study the internal activity of brain cells. Instead of adding genes and light, the lab technology known as designer receptors exclusively activated by designer drugs (DREADD) uses genes and chemicals to design a site on the neurons' surface called a receptor. DREADD works on a receptor that is coupled to a molecular switch inside a cell, called a G protein. These are tools to understand and manipulate the intracellular signaling system.

As was true in the case of fear and the use of optogenetics, appetite comes up again as a target for potential therapy, because so few current treatments have worked for patients. Attempts to understand the signaling mechanism could someday change that. Also, some of this work is about intracellular signaling systems that regulate food intake, providing potential insight into obesity and eating disorders. For example, compounds derived from cannabis, known as cannabinoids, are used to target receptor sites for appetite as well as for fear and other responses, to make animals eat faster. Interestingly, endocannabinoids are rapid-signaling systems in diverse regions.

Still other new ideas about modulating brain activity are coming up quickly, and many are remarkably ingenious. For example, scientists have used a virus to introduce a heat-sensitive calcium gene into the mouse midbrain. After a few weeks, they injected tiny magnetic particles into the same area and exposed the mice to an external magnetic field. They were then able to trace the activity of the neurons through their circuits. While optogenetics uses light, this technique combines magnets and heat to create a form of deep brain stimulation. Optogenetics is mainly oriented to networks of brain cells; DREADD is oriented to receptors that manage the cell's interior functions.

Another technique, called CLARITY, is helping scientists to learn about massive tissue connections. CLARITY strips away the fat from brains so that an intact tissue can be examined without the need to stain it. Like optogenetics, CLARITY involves light and the inspection of molecules but is mainly focused on the connectivity of nerve cells, helping to enable visualization of all kinds of connections across the brain. EEG, neuromodulators like TMS, and newer technologies like optogenetics, DREADD, and CLARITY all are workhorses of modern neuroscience, producing vast quantities of data on a massive scale. Taken together, neuroscientists have built brain data factories in laboratories all over the world.

TRANSLATING

Invention is one thing, innovation is another. Fifty years ago it was easy to envision flying cars scattered across the skies, because the technology was already on the shelf, but getting them into the production and transportation systems efficiently and safely, and monetizing their mass production, are other matters altogether. When physicians and scientists talk about "translational medicine," it's that challenge that they're referring to. It's not clear that any of the inventive technologies we've talked about in this chapter, other than tDCS, is ripe for fairly widespread use, and even direct current stimulation faces unsettled questions about its actual benefits.

From the point of view of conferring real value to its users, there's a good case to be made that deep brain stimulation should be first on the list for moving from experiments to common use. And there is a precedent for marketing medical devices that involve implanting an electrode array into the head: the cochlear implant, which was approved by the U.S. Food and Drug Administration, in the early 1980s, for patients with a certain kind of hearing loss.

However, DBS raises different concerns from cochlear implants or other implanted devices such as pacemakers or insulin pumps. DBS may change personalities in ways that are not well understood, perhaps even getting into potentially dangerous impulsive behaviors such as hypersexuality. Any device or drug that changes the brain raises sticky philosophical questions about informed consent, and these are clinically relevant. For example, is the recipient the same person who consented to the brain modification in the first place? (Of course, neurodegenerative diseases like dementia can raise similar questions, but we're talking here about deliberate interventions that could change personality.)

One ethical challenge that is becoming evident along with these applied technologies is the need to "design in" the ethics. For example, there is a risk that people who are dependent on neural implants will begin to wonder where the effects of the implant end and their "true selves" begin, similar to what some have reported with the use of medications like Prozac. These concerns might not only create their own psychological problems but also lead some in need to resist the use of these devices. Neuroengineers and neuroethicists need to work together to better understand patient reactions to these devices as they become part of medical care.

Then there are the practical problems related to going from invention to innovation. Some private companies that do experiments require participants to agree to allow the gadget to be removed at the end of the experiment, so their symptoms return. If researchers do allow a patient to keep a device, it might eventually stop working. If a patient has a device that wears out, or when the nearly inevitable scar tissue forms near the implanted electrodes, new parts will be needed. But it can be hard to find replacement parts, especially if the manufacturer goes out of business. And companies aren't required to share data from the devices that might be useful for research on making more and better ones (current DBS technology, for instance, is decades old), getting into issues about who owns and can profit from that information.

This is not a new story. Thomas Edison understood that it is one thing to invent a lightbulb, another matter to get it widely used. Nonetheless, no brain technology can properly be called an innovation unless it is accessible to those who want and need it.

6

SECURING

n the early nineteenth century, morphine was the first opiate isolated from a plant, the poppy, and was marketed by a German company called Merck. Almost immediately, it became so popular among soldiers that morphine addiction was known as the "army disease." One estimate is that four hundred thousand Americans were addicted after the Civil War. Much stronger than opium, morphine was called upon to treat just about any condition that involved pain, in the civilian world as well as the military. But during World War II, a few anesthesiologists noted that, under battlefield conditions, even men with terrible injuries reported little pain, though the same injuries would be agonizing in normal circumstances. There seemed to be something about the setting in which battlefield injuries occurred that caused them to be experienced differently than injuries in other settings, even though, physically, the injuries might identical. This experience led, in the 1950s, to great interest in the placebo effect.

Whatever opiate-like chemicals the brain itself makes to manage pain, morphine and other artificially administered opiates still have their place, when used appropriately. The trouble comes when they are "abused." We put that word in quotes because that

statement is a tautology—anything abusive is always troubling—
and the related advice to take all things in moderation is similarly
trivial. Those familiar counsels just kick the definitional problem down the road. Often "abuse" is a case-by-case judgment.
Take, for example, the use of drugs in war. In his remarkable
examination of the history of drug use in human conflict, Lukasz
Kamienski observes the ubiquity and antiquity of narcotics before,
during, and after people fight and attempt to kill one another.[1]
The weirdness and sheer terror of this situation as compared to
ordinary life, even for those who make a living as warfighters,
inevitably leads to a desire to modify consciousness accordingly. Think of "liquid courage," the term often used to describe
alcohol.

But the list goes far beyond a stiff drink. A brief accounting
includes opium among Homeric warriors, sixteenth-century
Turks, nineteenth-century Chinese, and many others; coca for
the Incas and nineteenth-century Indians; hashish for Napo-
leon's army in Egypt; marijuana for Americans in the Canal
Zone; amphetamine for British and German soldiers in World
War II and American soldiers in Vietnam; LSD for American,
British, and Czech soldiers during the Cold War; khat for child
soldiers in Asia and Africa; and on and on. Sometimes the drugs
are supplied by the armies, sometimes they are obtained by the
fighters themselves. Some military cultures actively promote
drug use and some are relatively "Calvinist" in their attitudes,
as Kamienski puts it; this would include the modern United
States. But that doesn't stop those on the front lines from finding ways to change their consciousness to deal with the bizarre
and traumatic conflict conditions, despite often terribly harm-
ful long-term consequences.

From managing pain to energizing for a fight to dealing with
the psychological and physical consequences that too often

follow, chemicals and combat are inseparable. There also are far more formalized ways that states seek to manage brains at war, a topic that might be called "neurosecurity." In a way, this is the oldest challenge for military commanders, who are always looking for an advantage over their adversary, and the human mind can be either the strongest or the weakest link in the chain.

Kamienski's list of narcotics doesn't include the chemicals our brains make to prepare for, address, and cope with intentional violence, whether motivated by a mugging or group solidarity or geopolitics. Nor does it include the ways in which the study of human behavior can inform and help shape the mind of the warrior. Apart from drugs and devices, the idea that psychology might make an important contribution to neurosecurity has been around for a long time, but it's been surprisingly hard to put into practice.

PSYCHOLOGY FOR THE FIGHTING MAN

Psychology for the Fighting Man was the title of a 1943 paperback that sold for twenty-five cents in civilian and military bookstores all over the world. The topics included morale, sight and perception, matching the soldier to the job, and racial differences. The book's popularity showed how far psychology had come since members of the new profession had to elbow their way into World War I. Then, American and British armies used the innovative and controversial "intelligence quotient," or IQ, to place people into specific jobs.

About twenty-five years later, the U.S. Army General Classification Test had a more ambitious role in determining which skills were "innate" and which ones could be trained into new servicemen and women. However, there were complaints that the potential for combat skills weren't being measured, especially physical qualities like strength and stamina. The air force set up its own Aviation Psychology Program, with an emphasis on the perceptual skills needed for flying aircraft, and it also employed statistics to measure hundreds of discrete aptitudes. By contrast, the predecessor to the Central Intelligence Agency, the Office for Strategic Services, developed a personality assessment program that placed potential recruits for intelligence work into situational tests, where they had to improvise responses to various challenging and frustrating situations. As the war finally drew to a close, thousands of psychologists participated in creating the Strategic Bombing Survey to determine what effect "strategic bombing" of German cities had had on civilian morale. Their answer: not much, except at the very end of the war, when the German people knew it had been lost.

After World War II, psychologists, who formerly did mostly animal experiments, started to work more with human beings as research subjects. They continued to study problems of training people for new tasks, as they had for the military effort, as well as looking into sensation and perception. That research easily morphed into weapons design, so that psychologists found common ground with physicists and physiologists. Concepts like "man–machine systems" and "human factors engineering" found their way into the lexicon of the 1950s.

Eventually, some of these ideas about engineering behavior veered off into B. F. Skinner's radical behaviorism, but many experimentalists focused more on the way that machines could be designed for human use. That tradition continues today, when neuroscientists work on brain–machine or brain–computer interfaces, though today the machine in question is often a computer, range finder, or dashboard. That idea, too, nearly ran off the rails, in the 1970s, when some DARPA officials became seriously interested in parapsychology, especially the idea of psychokinesis. At that time, Stanford Research Institute scientists were persuaded that the Israeli psychic Uri Geller could read thoughts, bend spoons, erase film, and move a compass needle, all with his thoughts alone. Ultimately, DARPA never funded a parapsychology project, though the army did, in the 1980s.

Another kind of mind reading, called biocybernetics, has proved a lot more productive than psychokinesis, and this is the kind we're familiar with now, though it has its roots in the early 1970s. The idea was that interaction between a computer and a brain would be based on the signals recorded by sensors via electroencephalography. Progress in that field has been slow but promising, as EEG caps are now a well-established technology. Even though the amplification of signals through the skull is a significant technical problem, it's clear that sufficient information can be obtained to improve perception. An example is the P300 wave, which is evoked almost instantaneously in the presence of a meaningful visual stimulus, such as a photograph of a familiar place, but before the participant is even aware of it.

The theories and tools of neurobiology might be used to improve the cognitive performance of security personnel who are faced with strenuous but sometimes mind-numbing tasks. An example is operators who stare at computer monitors for hours at a time. Their job often involves discerning potentially important clues coming from drones and other kinds of sensors. One DARPA system trains an operator wearing an EEG cap and looking at a busy video to consciously note when her brain has spontaneously emitted the telltale P300 brain wave, which is associated with recognition. In DARPA studies, an operator's "threat detection" capacity has been shown to be measurably improved. The trouble is that improvements in one sense so often turn out to be degrading in another. A drug that enhances working memory might inhibit the ability to filter out irrelevant stimuli.

Combining EEG with a video system is an example of another phenomenon that happens so quickly and effortlessly that we often don't notice it: convergence. Converging emerging technologies is an awkward phrase, but it describes something that happens with increasing frequency as several technologies can be combined to address what had once seemed to be intractable problems. A nice example from neurobiology is the problem of the blood–brain barrier, which works to protect us from pathogens but also makes it hard for some drugs to pass the highly selective gating mechanisms of the membrane. Some would say that the barrier is really a theoretical construct more than a physical object, because it involves a variety of different systems. In any case, it does present a challenge for drug delivery, one that might be circumvented through cleverly designed tiny machines produced by nanobiologists. Miniaturized, implantable devices could enable drugs to bypass the various constituents of the barrier. The hormone oxytocin does seem to reach targeted brain centers when it is introduced in quantity through the nasal passages. Nanobots carrying the stuff could be still more effective and might aid in procedures like interrogation.

Notions like these have more than a whiff of science fiction about them, but nanobot-emitting truth serum is only one of the convergent neurotechnologies theorized in a National Research Council report underwritten by the national intelligence community, which has a modest but sustained interest in the applications of neuroscience.[2] This interest is not only in the far-out possibilities for nano- and bio-convergence but also in closer-to-the-ground options, such as using neuroimaging devices to evaluate a job candidate's responses and to improve them as needed if they are ultimately admitted to training. That might be an advance on the primitive skill testing done now for all sorts of jobs, which can be compromised by factors like fatigue, boredom, or preoccupation with a domestic quarrel. Random, deliberately imposed stressors could be employed to capture actual operational experience and then determine how the participant's brain is processing and managing the stress.

Besides helping us to understand how the brain processes routine skills, these kinds of studies also might be applied to learning more about what is called fluid intelligence, the name given to creative problem-solving ability. Fluid intelligence is a critical personality attribute for effective leadership. Decision-making research suggests that a few distributed brain systems are responsible for processing beliefs that contribute to selecting appropriate or inappropriate actions. These brain systems can be monitored and potentially modulated during training or preparation exercises to optimize decision-making.

Among the tools available for characterizing individual decision-makers are longstanding ones, such as personality and emotional reactivity tests, and potential, emerging approaches in genetics, activation of neurohormones, and brain detectors. In practice, these last three technologies might render cumbersome screening mechanisms obsolete. For example, it seems plausible that special operations personnel who, during their training, must excel in various situational tests of decision-making under unexpected conditions are possessed of exceptionally fluid intelligence. In the opening scene of the *Star Trek* film *The Wrath of Khan*, Starfleet cadet James T. Kirk is given a no-win training test called the Kobayashi Maru. Kirk passes the test by secretly reprogramming the training simulator. In other words, he changed the rules of the game. If only the Starfleet Academy's twenty-third-century neuroscientists had been called in to examine the young Kirk for his fluid intelligence, the test itself might not have been necessary.

There are more routine concerns for commanders than out-smarting war games, like sleep deprivation and fatigue, which it is commonly agreed account for a distressing number of combat accidents, including attacks against one's own forces. Traditionally, when it's not possible for warfighters to get more than a few hours of sleep, they depend on nutrition, especially carbohydrates and caffeine. (On the same physiological principle, sleep-deprived travelers often find themselves eating more than usual.) Combining both carbs and caffeine, the U.S. military's little-loved Meals Ready to Eat often include caffeinated chewing gum. There are pharmaceutical options, such as plant-derived phytochemicals that have some nutritional value and that function as antioxidants and anti-inflammatories to promote brain health, not to mention brain chemicals like adenosine, ammonia, and dopamine to treat fatigue.

SOCIAL NEUROSCIENCE

The "softer" side of neuroscience—psychology and allied social science fields—has a murkier record. One long-sought goal has been to provide government with predictions of unfolding crises, which would involve massive data gathering. In 1964, long before many people became aware of "big data," at a time when America's major national security problem was communist insurgent movements in Latin America, Africa, and Asia, a group of U.S. Army social scientists started Project Camelot. As described by historian Joy Rohde, the goal was to identify the social psychological dynamics in countries like Bolivia and Colombia that

American "counterinsurgency" efforts would need to take into account in anticipating social change.[3] Project Camelot was more than a think tank of social scientists; a key element was to assemble variables that could computerize predictions of revolution. But the Pentagon canceled the project in 1965, after the Chilean government complained to the United States that a Chilean Camelot consultant had conducted unapproved interviews with social scientists in the country without disclosing his Defense Department connection. This was perhaps the first time that social science was called an imperialist invasion.

In 1967, the Advanced Research Projects Agency (DARPA's predecessor) funded the World Event/Interaction Survey, which was supposed to crunch information about the international situation and spew out predictions for policy makers. Nearly forty years later, a similar project, called Total Information Awareness, also ran into political trouble. The project was directed at identifying terrorist planning, but its computer simulations required collecting data about every possible kind of online information, including e-mails and personal communications, which appeared to invade the privacy of average citizens.

It's easy to understand why security planners would love a predictive system, but all such efforts have run aground at two points: how to input "raw" information about some incident (because no information is ever wholly value free) and how to convert the data into an actionable forecast. A legion of experts on the Middle East failed to predict the rise of the Islamic State, much less recent, low-tech terror attacks. A generation of Sovietologists did not foresee that street protests would lead to the collapse of the Berlin Wall and finally of the Soviet empire, nor did Arabists predict the Arab Spring, nor did international observers anticipate the outbreak of World War I. Human intelligence and vast quantities of metadata seem ill-equipped to help forecast geopolitical events at this scale. Despite decades of sophisticated political science and game theorizing, the truth is that we are very bad at anticipating great historic turns.

What is more susceptible to research than these big geopolitical questions is the way in which the brain is influenced by social networks. Over the past twenty-five years, basic biological sciences have acquired new tools that have helped to provide compelling foundational knowledge about the brain and behavior, knowledge that can be combined with social science research. As noted in the work of neuroscientists like Nicholas Christakis, for example, social network proximity is a predictor of obesity.[4] Emily Falk notes that people who connect people to others have different brain activity from those whose networks are more confined.[5] The transmission of information through social networks connects brain dynamics to social dynamics.

Genetics is yet another level of analysis. New technologies are producing wiring diagrams of the brain and examining how neural systems are produced and regulated by perhaps eight thousand genes. Using scans, computers, and sophisticated algorithms, there are now many ways to examine brain activity in the laboratory under a variety of stimulus conditions. These studies of basic biology are being linked back to social psychology, such as by looking at the ways in which different cultures interpret facial expressions and how those interpretations are reflected in brain activity. An international research team found that groups of people in Japan and the United States had significant differences in neurological responses to images of faces that showed fear, differences that registered in the amygdala.[6] Related work is being done on empathy for those in one's own racial group as compared to others. Activity in the medial prefrontal cortex, which seems to play an important role in complex thought, is associated with self-reports of empathy when viewing pictures of distressed people thought to be members of one's own racial group.

Also being studied is whether these responses are at least partly mediated by the large blocks of inherited genes that help to define each ethnicity. However, these kinds of in-group/out-group biases are not simply built into our genes. They also result from developmental processes of environmental interaction with genetics, studied in a field called epigenetics, and therefore these attitudes can be studied and modified. MIT researchers have found some evidence that groups with less perceived power can change their attitudes toward those with seen as dominating them (such as Palestinians' view of Israelis) if they are given the opportunity to speak and be listened to, though it's not clear how long that effect lasts.

The ultimate applications of such knowledge would have to take place outside the lab, so it is important to associate such data with the ways in which people actually understand the world. To that end, one ambitious Defense Department project aims to learn how stories influence political radicalization and the way in which groups are moved by narratives. Stories we tell ourselves about our purpose and place in the world follow a kind of logic and help to explain how people make moral judgments. The same processes take place in the minds of terrorists.

At the individual level, it seems that neurobiology should be able to tell us something. It could be important to know that, before an important negotiation, Vladimir Putin had a massage that increased his production of the "trust hormone" oxytocin. However, it's a big step from Putin's massage to anticipating something like a terrorist attack, much less a historic geopolitical event. Even if it were attainable, that kind of knowledge would present hazardous prospects of its own. A frankly repressive regime could anticipate and suppress social developments that could someday lead to a morally justified insurrection. But any futuristic forecasting system that integrates knowledge from basic biology with social psychology still would produce only predictive models with large margins of error. Anyone looking for laws of future history would do best to consult the writings of Isaac Asimov, for that idea will remain the province of science fiction.

What is not fictional is that governments will continue to seek any strategic advantages that science might enable them to achieve, and advances in neurobiology do present fascinating opportunities that can't be ignored. Unfortunately, as has happened more than once, even benign governments could overinterpret forecasts provided by those who are regarded as "experts," especially if they clothe their predictions in fancy science. Less admirable states could use them as excuses for preventive first-strikes on adversaries. Any emerging technology is subject to the Collingridge dilemma; it can't be controlled until it has been developed and disseminated, by which time it may be too late.

THE THIRD OFFSET

The issues confronting programs like Project Camelot and Total Information Awareness weren't only political. More basically, they encountered political resistance because of ethical questions. A new U.S. strategic doctrine called the Third Offset also poses an important ethical challenge for neuroscientists. The ethical issues related to neuroscience and national security were not among the topics discussed at the Dana Foundation's landmark neuroethics conference, "Mapping the Field," in 2002. But only a year after the Dana conference, *Nature* published a tough editorial, "The Silence of the Neuroengineers."[7] The editors accused Pentagon-funded investigators of failing to respond to, or even consider, questions about the potential uses of technologies like brain–machine interfaces. An indignant letter from the chief scientist at the Defense Advanced Research Projects Agency suggested that the *Nature* editors harbored a prejudicial attitude, failing to take into account the medical advances that could eventuate from DARPA-funded neuroscience. Since then, the possible military and intelligence application of modern neurotechnologies has stimulated a modest literature. Nonetheless, the field is still underperforming in its attention to the national security environment.

The arguments for intensifying a focus on neurosecurity are many, including a steady pattern of substantial funding, in the tens of millions of dollars, for neuroscience projects by various national security agencies. Though DARPA has received the most attention among those who have followed these developments, both the Intelligence Advanced Research Projects Activity and the Office of Naval Research have substantial programs in fields like electromagnetic neurostimulation and computational neuroscience. In addition, the dual use argument that reverberated in the exchange about the *Nature* editorial points to the momentum behind federal funding for neuroscience. Even those who worry about "militarized" science are put in the awkward position of threading a moral needle when, for example, new prosthetics for severely incapacitated persons are in the offing and when new therapies for dementia and trauma are so desperately needed. Such is the case in the U.S. BRAIN initiative, in which DARPA plays a key role.

For example, intelligence collectors are those who obtain data from various sources and pass it on to intelligence analysts. Improving detection and interpretation of raw data is of great interest to intelligence collectors. The U.S. Department of Defense is experimenting with EEG signals to help train intelligence analysts to pay more conscious attention to potential threats, using visual information from cameras on drone aircraft. Analysts will have to learn to perceive those visual cues better, through response to some kind of change in their environment, such as visual, auditory, or haptic (touch) signals. This idea dates back to the concept of biofeedback, from the 1970s.

Though the relevant technologies are vastly improved over what they were during the Cold War, neurosecurity is not a new concern for defense planners. Since antiquity, commanders have striven to achieve psychological advantages over adversaries, advantages that ranged from propaganda to narcotics. The two world wars saw the introduction of intelligence tests and personality inventories. Cold War intelligence and military officials worried about whether a new compound called LSD-25 could be used as a "truth serum" or a way to demoralize fighters. As strange as these concerns might seem to us now, still more bizarre were serious explorations of extrasensory perception, including remote viewing and telekinesis.

One important lesson learned from American war-fighting capacity during World War II was that although U.S. industrial might outproduced all the other protagonists in terms of sheer quantity, the quality of war materials often lagged behind that of Nazi Germany and Imperial Japan. Therefore, since the Truman administration, a consistent premise of U.S. policy has been to focus on technological superiority over all actual and potential adversaries. This posture has virtually assured that even implausible "technologies" that might confer an advantage will be considered. It also has resonated well with a country that has been transformed into a post–World War II national security state in which essentially all sectors play a role in the defense of the nation and all societal purposes and resources may be subordinated to national security goals.

The appreciation of the importance of a technological edge has been characterized among U.S. defense planners as an "offset strategy." For those strategists, nuclear weapons represented the first offset in the face of a Soviet enemy with significant numerical advantages in conventional weapons. The doctrine of overwhelming nuclear capacity on both the U.S. and Soviet sides was called mutual assured destruction, and for all of its MAD quality, the strategy enabled a balance of power during the Cold War and was subject to a more-or-less successful nonproliferation regime. The second offset included precision-guided munitions such as laser-guided "smart bombs" and computerized command-and-control systems, which proved themselves in the Gulf War of 1990–1991.

These technologies were clearly cutting edge in their day, but new possibilities have emerged that require new ways of thinking about defense research and development. Moreover, national security strategists face a multipolar world that also includes nonstate actors capable of terror attacks that pose mainly a psychological rather than an existential threat. The result is a moving target in several senses of the term, one that requires exploration of novel technological approaches.

For several years, these new technologies have been collected under the heading of the Third Offset strategy, described by Real Clear Defense as "an attempt to offset shrinking U.S. military force structure and declining technological superiority in an era of great power competition—a challenge that military leaders have not grappled with in at least a generation."[8] The precise outlines of third offset technologies aren't as clear as in the first two offsets, so observers wait for Pentagon budgets to be released in order to learn of their research and development commitments.

There are of course classified, or "black," projects that are not publicly stated. Nonetheless, some of the public budget specifics are of particular interest. One budget line calls for autonomous deep learning machines and systems for early warning based on crunching big data. The idea behind deep learning is that a machine could be programmed with algorithms that enable it to create abstract representations of data that can be applied to tasks like speech recognition, natural language development, and even customer relations. Some of the work on deep learning is based on lessons from the way the brain processes and codes information.

Deep learning also would contribute to other Pentagon goals, including enabling human–machine collaboration to help human operators make decisions and performing assisted-human operations so that humans can operate more efficiently with the help of machines like exoskeletons. Today, there are only glimmers of how efficient a human–machine team could be if the human could interact with the machine without having to push a button or move a joystick. That would require an advanced interface, which is of interest to industry, for manufacturing purposes, as well as the military, and ultimately for everyday use. These would be vastly more powerful and natural-feeling than the voice-controlled home electronic appliances that are currently on the market. Such human–machine teaming would require the two to communicate with each other in order to integrate into a seamless whole.

DARPA doesn't have much to say about this work in public, but one concern about the practicality of any new human–machine interface is presumably its susceptibility to hacking by hostile forces. Semiautonomous weapons that can be "hardened" for cyber warfare will have a role in any prospective system. Much of the rest of this chapter will circle back to some of these ideas.

BRAIN TO BRAIN

Decades after Clint Eastwood's character bonded with his fighter plane in *Firefox,* Tony Stark became Iron Man on screen, through his advanced human–machine interface. Such intimate human–machine teaming is pretty far off, but in the near term we will see robotic assistants like prosthetics and exoskeletons. Already, computers help analyze satellite images to give operators a technological edge, and in the civilian world, self-driving cars are practicable if not yet socially acceptable. Getting closer to Iron Man will require not only "hard" engineering advances but also "soft" psychological engineering, concerned with human factors in working with ever more sophisticated machines, and a field called neuroergonomics, which applies findings from neuroscience to help understand the human side of human–machine teaming.

Because the third offset technologies are not clearly definable, one needs to be alert to other projects that fit within the scope of the new doctrine, complement some of its goals, and fill out the neuroethical issues. Naturally, many of these projects are dual use. An example is DARPA's Systems-Based Neurotechnology for Emerging Therapies program. In cooperation with other federal agencies and private industry, DARPA aims to develop improved electronic microarray chips for deep brain stimulation to treat neuropsychiatric disorders and to be used as part of prosthetic systems. These would be closed-loop devices that both monitor and regulate brain functions. Implantable chips are envisioned to go well beyond the standard ninety-six-electrode array. They would be advantageous not only for new therapies but also for enhanced performance of a variety of tasks.

An open question is whether material might be developed to create biocompatible microarrays of thousands or even tens of thousands of electrodes. DARPA's Neural Engineering System Design project "aims to develop an implantable neural interface able to provide unprecedented signal resolution and data transfer bandwidth between the brain and electronics. . . . The goal is to achieve this communications link in a biocompatible device no larger than one cubic centimeter in size, roughly the volume of two nickels stacked back to back."[9]

A reliable human–machine interface would be a critical component of the third offset. But some DARPA planners wonder whether such a device could turn out to be still more powerful, enabling direct human-to-human communication over distances. The controversial *Nature* editorial of 2003 anticipated these efforts in its indictment of passive neuroscientists who were failing to address the underlying ethical questions: "The agency wants to create systems that could relay messages, such as images and sounds, between human brains and machines, or even from human to human."

One primitive experiment proving that the concept of brain-to-brain communication could be realized took place in 2013 at the University of Washington.[10] One researcher, wearing an EEG cap, thought about moving a finger to hit a target on a computer screen. That signal was sent over the internet to another scientist, who was wearing a magnetic stimulation cap. Without knowing why, the second researcher moved his finger to hit the space bar on a keyboard, likening the sensation to a nervous tic. It was a provocative exercise, but obviously an advanced bioelectronic chip implant could go much further. The UW team has continued its work, with two people networked to play a game similar to Tetris.

Still more remarkable would be a system that not only would enable instant communication between brains but also would allow a network of brains to apply their processing power to a specific problem. Imagine that a financial crisis was threatening international markets. Some number of financial experts could be brought together in a group neural interface and could cooperate to find a solution, each applying his or her unique expertise, without the encumbrances of speech, personality differences, or the ego needs of a lot of smart people. An analogous scenario could apply to a national security crisis in which the collective and unfettered experience and intellect of military and diplomatic experts could be joined in a neural net to sort out the issues and recommend options. Individual differences that were irrelevant to the problem at hand could, in theory, be filtered out. Soldiers in a battle space or workers on an oil rig are other examples of cases in which directly networked brains could operate on a challenging situation so as to complement each other's knowledge and skills, a kind of brain swarm.

Some proof-of-concept experiments with other species suggest that the idea of a collective networked brain, as intimated in the *Star Trek* group the Borg, isn't such a bizarre fantasy. A Duke University–University of Pennsylvania team linked the brains of several rats—and, in another experiment, several monkeys—in a "brain net" that enabled them to combine their brain power to solve problems more efficiently than they could as individuals. In the monkey experiment, electrical activity was recorded from the brains of two monkeys in separate rooms.[11] Each had 50 percent control of a virtual arm on a computer screen in front of them. Together, they moved the arm to receive a reward, either using a joystick or in response to electrical modulations in each monkey's brain. It took quite a bit of training, but the better the monkeys focused on the avatar arm, as measured by correlations of the neurons that were firing in each brain, they better they performed in moving the arm around. Multiply the numbers in a task and you reduce the chance that fatigue or loss of a participating brain would impair the project. The experiment's team leader, Miguel Nicolelis, observed, "Even if one monkey dropped out in one trial, the brain net is resilient. Imagine if you had, not three, but a million. That would be extremely resilient."[12]

This experiment keyed into a few hundred neurons in the motor control centers of each monkey. It wasn't a complex task, but the results still were impressive. Could problems that call on higher-level cognition, like financial decisions or combat tactics, be subjected to a neural swarm? It might be that, at that level of operations, the human brain presents obstacles that no such system can solve. Furthermore, no one can say how each monkey "felt" while it was implicitly cooperating with another monkey's brain to move the virtual arm. As humans, we can't say what they experienced while in the brain net, or how it would feel to us. Would the others have insight into our most intimate thoughts? Nonetheless, at an empirical level it would be interesting, and potentially important for human therapies, if one linked brain could assist in healing an injured brain. Stroke patients with language problems due to aphasia have physical manifestations of the damage in distinct areas. Perhaps networking with healthy brains could help them find alternative pathways in the speech area to improve their function.

A brain scientist involved in brain-to-brain research will tell you that there are no insurmountable obstacles, even to shared higher-level cognition, so long as you can find the mechanical keys to the process. Besides the engineering problems, however, experiments to investigate what sorts of human-to-human brain linking might be possible will encounter substantial ethical obstacles, especially if the technology involves invasive neural implants. Imagine, however, that two Parkinson's disease patients were coincidentally scheduled to have surgery at the same time for the placement of electrodes to help control their motor symptoms. Individual Parkinson's patients have already volunteered for experiments involving auditory pathways while undergoing therapeutic surgery, so that in itself wouldn't be novel. In a brain net experiment, two such patients could be asked to give informed consent to be linked during the surgery, to engage in a variety of motor and intellectual tasks, and to report on the phenomenology of the experience. That would be a fascinating proof of concept. But the ethical problem of linking a patient with a brain disorder to a human volunteer with an implant would be hard to surmount. The solution might rest on some kind of external brain net device, like an EEG cap.

RETURN OF THE CENTAURS

There are other technologies relevant to the third offset strategy with a human aspect. One of them is called "centaurs." This concept goes far beyond militarized exoskeletons or mechanical dogs that carry weights alongside the warfighter in the field. The centaur concept involves human–machine teaming systems in which the machine and the human are joined to combine their special skills. Machines are precise and reliable but inflexible and unspontaneous, while humans are flexible and imaginative but often imprecise. So far at least, machines can't think outside of the box they are in (though deep learning systems based in part on neural networks may change that), while humans are subject to bias and fatigue. But these "cognitive teams" will enable humans to do their jobs faster and more accurately.

There is already an example of an airborne centaur, the F-35 Lightning II. The F-35 receives vast quantities of data, which it analyzes and displays it to the pilot. The goal is to enable the pilot to make better decisions. Obviously, much depends on both the algorithms the onboard computers use and the skill of the pilot in making the most of the information provided. Basic neuroscience will be and already is part of both processes, in the design of the software for maximum reliability and the way in which the information is conveyed to the human so that it can be digested and acted upon without an overwhelming data deluge.

For neurotechnology, the next question is whether there could be implants that create an instantaneous human–aircraft connection. Setting aside the inherent risks of brain surgery, we note that such projects are under way, though still in their infancy. Neural Engineering Systems Design involves an implantable technology that would enable signal resolution and data transfer between the human brain and the digital world. Reliable Neural-Interface Technology is an implant that proposes to extract information from the nervous system that would be needed for a human operator to effectively control an exoskeleton or to interface with other complex technologies. Iron Man Tony Stark is not yet with us, but he might not be so far away.

A NEW ARRAY

Putting any reliable system like this into a routine, real-world situation would require systems that we can barely imagine today. If you can suspend your disbelief, you can understand why military planners fantasize about taking an idea like this into the field so that warfighters can instantly transmit information to one another in a kind of telepathy. If that notion stretches your imagination, consider the possibility that, through some kind of direct brain-to-brain network, the knowledge bases of experts from various fields of learning could be pooled so that any one of them would have intellectual access to the information that it took each of them years to acquire in their separate disciplines.

Brain–machine and brain–brain communications systems are examples of the convergence of neuroscience and engineering. People with serious diseases like Parkinson's, amyotrophic lateral sclerosis, or quadriplegia could be helped by a still experimental system called BrainGate, which was first developed at Brown University. BrainGate uses a one-hundred-microelectrode array implanted in the brain to track neural activity and translate it into commands executed by a computer, performing tasks such as controlling the lights and temperature in a room or moving a cursor over a screen.

Still another version of convergence is on the radar of those in military planning who are responsible for investments in neuroscience. They are among those who talk about a new field they call "engineering biology," using biological systems to do things that human-made machines can't do. Using tools from various fields, including genomics, imaging devices, and information science, the ultimate goal is to create biological systems that can be programmed to do things like make new vaccines and medications or run new manufacturing processes for just about any purpose. Neurotechnology applications are at the top of this list, such as creating for amputees new prosthetic limbs that are themselves partly biological and "smart," and no doubt wireless, rather than the inert artificial arms and legs now being developed, which depend on wiring to muscle and nerve groups. Though the present iterations are very sophisticated, they couldn't compete with new prostheses built on biological systems rather than on plastics and metals.

PROFESSIONAL ETHICS

In the wake of objections from many employees, in 2018, Google withdrew from a Pentagon contract in support of Project Maven, which employs software to support the analysis of drone imagery. Part of the Department of Defense's response to such Silicon Valley concerns is a new Joint Artificial Intelligence Center, which includes a focus on "ethics, humanitarian considerations, and both short-term and long-term AI safety."[13] Drone technology is part of a larger movement that is enabling technologies, including offensive weapons systems, to become more autonomous. The concept of neural networks, which has been explored and expanded upon by neuroscientists, is providing some interesting options for making software that can possess at least some degree of an appreciation of context when running those systems. There is a raging debate within the defense community about the extent to which any weapons system should be allowed to make a determination on its own about engaging in a lethal attack. The problem is that, despite a warrior culture that resists giving over such power to machines, the tempo of the modern battlefield is so rapid, and the data being collected by sensors and computers is so complex, that commanders are being drawn inexorably toward a more automated approach to war fighting.

The ethical tension in the relationships among science, engineering, and war fighting isn't new. The U.S. National Academies have their origins in President Lincoln's need to screen new war-fighting technologies. Albert Einstein regretted his role in encouraging President Roosevelt to create the Manhattan Project. Defense funding of university-based research was a flash point in antiwar protests in the late 1960s. Psychiatrists and psychologists assessed intelligence, personality, and small-group dynamics during the world wars and in soldiers' and Marines' responses to the atomic field tests in the 1950s. Social scientists were sought to help understand communist subversive movements in the developing world during the 1960s. Modern U.S. military superiority would be literally unthinkable without the massive financing of the academic world since Sputnik. The *Nature* editorial "The Silence of the Neuroengineers" urged neuroscientists to be more aware of the implications of work that might be funded by agencies like DARPA, sparking an angry response by a senior member of the Defense Sciences Office, who noted the benefits of their work to civilians, including in medical care.[14]

We believe there are ways for individuals and organizations to think through these seemingly conflicting responsibilities. In a June 2018 op-ed in *The Atlantic*, former secretary of state Henry Kissinger called on technologists and humanists to join together in leading the way toward a philosophical framework for the ethically challenging new era of neuroscience and its military applications.[15] This is a fine goal, but it is made far more complex with the introduction of defense planners into the conversation. A first step is trust. Academics may be surprised to learn that ethics is written in the DNA of our military culture, whereas military planners must put aside any stereotypes about pointy-headed radical peaceniks run amok on campus. (That doesn't mean there aren't exceptions in both cases.)

From the technologists' side, at first, the problem seems daunting. How are we to reconcile personal ethics, a desire to expand knowledge and contribute to human flourishing, and the relentless demands of national security? All involve the nature of responsibility and the realities of an increasingly competitive and often violent world, one in which the international arrangements that have prevented global catastrophe for seventy years seem now to be more under stress than ever. Most of all, scientists and engineers don't want to be in the position satirized in 1960s comedian Tom Lehrer's ditty, "'Once the rockets are up, who cares where they come down? / That's not my department,' says Wernher von Braun."

We suggest a framework based on the principles of proximity and engagement. Proximity refers to the known role that one's own work would play in causing death or injury. The more proximate the role, the more reasonable it is to ask questions about how the use of the technology, such as how it would be governed by the laws of armed conflict and command structures. Admittedly, basic science also can end up in applications with unintended effects. Those can't always be gamed out in advance, but scenarios can at least sometimes be imagined, and they, too, can raise appropriate questions.

Another aspect of proximity is errors of omission: What harm will be done if I don't undertake this work? The latter question leads to the principle of engagement. As well as their role in the work, scientists and engineers need to consider the consequences of their conscientious absence from a conversation. If they don't insist on building acceptable and verifiable safeguards into a system, someone else will, and not necessarily in a form they would endorse. To have a voice at the table, you need to have a seat at the table.

7

HEALING

The goal of neuroscience is not only to understand the brain but also to heal it. That is the underlying rationale for both public funding and private investment. The brain afflicted with mental illness is still in a context, like all brains. But how should the context be interpreted? Modern neuroscience has spawned a fierce debate between those who continue to favor a symptom-based categorization of mental illness and those who want to move toward a system based more on neuroscience theory. It's hard to deny that certain diagnostic categories have emerged from cultural prejudices. One example is homosexuality, which was dropped from the list of mental illnesses published by the American Psychiatric Association, in the 1970s, after much controversy. Many neuroscientists argue that a more objective, science-based approach is needed and that neuroscience is approaching a level of maturity sufficient to support that approach.

The Hippocratic mandate of medicine is to reduce and ameliorate human suffering. Even without a deep theory of the way the brain works, current psychotherapies can help manage panic attacks, chronic obsessions, thought disorders, disruptive attentional capabilities, and excessive mood swings. But we can do far better, and we need to. Promoting neurogenesis (growth of new neurons) has been suggested as a defense against vulnerability to relapse into depression. And neurogenesis is revealed by pharmacology (Prozac) or by physical activity, and perhaps also by cognitive therapy.

And sometimes the treatments are surprising. The typical treatments for major depression usually take close to six weeks. That is not very helpful to someone in the grip of the anguish of this serious illness, which includes unrelenting anxiety and sorrow, with little relief and little sense of hope, and suicidal ideation that is not so distant and is without amelioration. Recent findings indicate that ketamine, which is known to block N-methyl-D-aspartate receptors in the brain and to influence the neurotransmitter glutamate, can, within hours, provide relief to patients suffering from depression. How long the effect will last is not known.

SERIOUS MENTAL ILLNESS

Nonetheless, progress in treating serious mental illnesses like schizophrenia has been slow. Several groups have uncovered vulnerability genes and, more recently, epigenetic changes that further facilitate the devolution of capability in these individuals. For instance, Daniel Weinberger's group at the Lieber Institute for Brain Development has pinpointed a number of genes that increase the risk of schizophrenia. Their studies of thousands of postmortem brains are among those that apply new and emerging neuroscience to diseases like psychotic disorders that have been frustratingly resistant to treatment.

As sentiment turned against lobotomy and electroconvulsive therapy in the 1950s, there was a turn toward antipsychotic drugs

that block dopamine receptors, ameliorating the most disturb-
ing behavior and making it possible to think of removing patients
from long-term hospitalization. That movement, called deinstitutionalization, proceeded apace—though without adequate
arrangements in the community for many people who still need
intensive management—employing the atypical antipsychotics
that avoid the worst side effects of the older drugs. Since the early
2000s, the focus has been on the N-methyl-D-aspartate receptor for GABA, which regulates fear and anxiety. Glutamate
modulation might be the future of the treatment of schizophrenia.
Glutamate is an excitatory neurotransmitter; GABA, another
fundamental neurotransmitter, is an inhibitory neurotransmitters. Their balance in the brain is critical. More importantly, it
might augur a more valid theory-based approach to treatment.

Addictions are another important target for neuroscience-
based therapies. These are a kind of obsessive-compulsive disorder directed toward various objects, from sweet drinks to sex to
games of chance to drugs. Addictions in all forms involve an
obsessive, destructive loss of control. They might be called the
rape of the will. The neuroscience of addiction suggests three
phrases: binge and intoxication, withdrawal and negative affect,
and increasing sensitivity to pain and emotional distress, perhaps
involving regions of the lateral amygdala and the stria terminalis.
Neurotransmitters such as dopamine were once thought to be
more centrally involved in addictions than it is presently believed,
but they certainly are critical in the reward pathways that are
associated with incentives to procure the addiction-focused object.
Experiments have demonstrated that corticotropin-releasing
hormone neurons in the amygdala or the nucleus accumbens
focus and enhance the motivation for natural sweets and psycho-
tropic drugs such as cocaine. Other structures, such as the bed

nucleus of the stria terminalis, may worsen the effects of withdrawal from drugs.

Little is known about how environment is related to obsessive behaviors or descent into pervasive negative thoughts and emotions. Addiction is usually anything but a fun experience. It involves a chronic chase after an object, which can never be fully satisfied, a depletion of resources, and increasingly desperate and negative emotions. In all cases, addiction is about a compulsive narrowing of choice. With the growth of, let's say, opiate addiction, the euphoric states dissipate and the compulsions increase. We know something about the molecular mechanisms (such as the cAMP response element binding protein) that underlies the regulation of genes, and about targeting synaptic functioning, the intracellular signaling systems that underlie addiction, tolerance, sensitization, dependence, and withdrawal. But there are no magic cures. For many years addiction has been construed as a "disease" of the brain. But social context is overwhelming.

Indeed, the current climate of addiction, particularly to opiates, and the tragic epidemic of overdose across every demographic category in the United States, is testimony to the climate of addiction. We do know how to recover individuals once they have overdosed (through the use of naltrexone) or to block cravings for opiate by using blockers such as methadone. We know, by using neuroimaging techniques, that diverse regions of the brain are differentiated by what are now called "substance abuse disorders." These include critical regions of the neocortex, such as the prefrontal and cingulate regions, and subcortical regions such as the amygdala and the bed nucleus of the stria terminalis and nucleus accumbens. Avoidance is still the best therapy to prevent vulnerability, but a brain wired for substance abuse is like saying "Just say no" when the obsessive compulsive vulnerability is to procure and consume. And, for a subset, there is what George Koob and his colleagues have called the "dark side"—addiction comorbid with depression, with existential angst and discomfort, and with the sense of doom.[1] William James depicted this state of the brain in the battered or "sick soul."[2]

Controversy has dogged brain-based novel therapies. For example, in the 1940s, it was noticed that people experiencing the stress of mental illness might be made less excitable if the connections between awareness and sensation could be modified. That was the idea behind a surgical procedure called transorbital lobotomy, so called because it involved using a surgical instrument to work past the eye, back to the fibers that connect the brain's gray matter to the thalamus, which helps to regulate consciousness. (If that sounds grisly, it was considered an advance over the previous technique, which involved boring holes in the skull.) In the mid-twentieth century, cutting those fibers seemed like a rational and anatomy-based alternative to the various shock therapies that were then in use in asylums. At first celebrated as a miracle cure—the Portuguese pioneer of the technique won a Nobel Prize—psychiatrists and the media came to recognize that many of these "psychosurgeries" were going terribly wrong. It was becoming clear that, in spite of good intentions, there was more to treating mental illness than simply making an incision.

At around the same time, a technique used to stun pigs in the slaughterhouse was applied to patients with mental illness, deliberately inducing a seizure that seemed to lead to remission of the disease. The new electroconvulsive therapy was both rapidly embraced by some and disdained by others. Many Americans learned about the benefits of ECT at the movies (*Snake Pit*, in 1948), but within thirty years, films turned against it (*One Flew Over the Cuckoo's Nest*, in 1975, though the mental hospital setting was incidental to the allegory of dictatorship). Treating mental illness wasn't as simple a matter as stopping a charging bull in his tracks. Nonetheless, despite all the controversy, modern ECT can be a relatively safe and effective last line of treatment for people with depression who haven't responded to talk and drug therapies.

ELECTRIC THERAPIES

The idea of treating ailments with electricity dates back to antiquity. Around 50 BC, Scribonius Largus, the physician to the Roman emperor Claudius, advised that "to immediately remove and permanently cure a headache, however long-lasting and intolerable, a live black torpedo [electric eel] is put on the place which is in pain, until the pain ceases and the part grows numb."[3] The first attempt to stimulate the brain through the skull occurred in 1755, when Charles Le Roy tried to cure a twenty-one-year-old man of his blindness by applying electrical impulses to his head. The impulses generated phosphenes (glowing spots) on the patient's retina but were not strong enough to affect the brain. Although Le Roy failed to cure blindness, he proved that nervous tissue responds to electricity. His work, along with the study of electromagnetism, raised the possibility that strong magnetic fields could stimulate brain tissue.

In 1831, the renowned English chemist and physicist Michael Faraday demonstrated that when an electric current is passed through a coil of wire, the fluctuating magnetic field generated around the primary coil will induce a current in a second, neighboring coil. By the beginning of the twentieth century, scientists had established that magnetic fields could modify neural activity, but it was not yet possible to generate large electrical

currents using a magnet. In 1980, the British neurophysiologists
P. A. Merton and H. B. Morton reported stimulating the cerebral cortex of an intact, living human participant with "brief but
very high-voltage shocks . . . without undue discomfort."[4]

In 1985, Anthony Barker and colleagues at the University of
Sheffield in England succeeded in using transcranial electromagnetic induction to stimulate the human motor cortex, thereby
inventing transcranial magnetic stimulation.[5] In some ways,
TMS is similar to functional magnetic resonance imaging (fMRI)
in that both employ intense magnetic fields, but there are important differences. First, whereas fMRI is an imaging technique,
TMS is a stimulation and therapeutic technique. Second, the
magnetic field plays a central role in fMRI, while the induced
electrical current is paramount in TMS. Third, fMRI machines
are expensive and bulky and require extensive technical knowledge for safe and effective operation, but TMS equipment is much
smaller and easily can be used in a doctor's office.

As a therapeutic tool, repetitive TMS can be customized
to treat different illnesses. In most stroke victims, for example,
one brain hemisphere is retarded while the other is largely unaffected. This asymmetry decreases motor cortex activity in the
affected hemisphere and increases activity in the unaffected
hemisphere. Restoring function in both hemispheres is essential
if the stroke victim is to recover. Using localized magnetic pulses,
repetitive TMS can help to balance the neurological activity of
the two hemispheres by enhancing the excitability of the cortical
neurons on the injured side of the brain and suppressing activity
on the unaffected side.

Repetitive TMS is often compared to electroconvulsive therapy, which delivers a direct electrical shock to the brain rather
than a magnetic pulse. ECT, developed long before TMS, is still
the standard treatment for adults suffering from treatment-
resistant major depressive disorder, but there are indications that

this situation could soon change. In 2007, the U.S. Food and Drug Administration convened an expert panel to determine the risk–benefit profile of the NeuroStar Advanced Therapy System (a transcranial magnetic stimulator manufactured by Neuronetics), compared to standard ECT therapy. In a letter to the FDA panel, a psychiatrist stated that, in her practice, ECT produced lingering side effects: patients were unable to work or drive for two or three weeks after an ECT session, and individuals of modest means could not afford to take leave for that amount of time. The psychiatrist found that repetitive TMS produced the same therapeutic benefit as ECT with fewer long-term side effects. TMS might be the successor to ECT for the treatment of severe intractable depression and might prove to be superior. It is noninvasive and can be used as an outpatient procedure with no significant side effects. In 2008, the FDA approved a TMS device for the treatment of major depression in adults for whom medication hasn't worked.

Originally developed in the 1980s as a diagnostic aid for neurologists, TMS has since helped to map brain circuitry and connectivity, and it offers therapeutic benefits as well. To perform the technique, a technician holds an iron-core insulated coil embedded in plastic on one side of a patient's head while a large, brief current is passed through the coil. The current generates a magnetic pulse that painlessly penetrates the layers of skin, muscle, and bone covering the brain and induces weak, localized electrical currents in the cerebral cortex. Although the mechanisms by which the localized currents modulate neuronal activity are not fully understood, it is believed that the induced electrical field triggers the flow of ions across neuronal membranes and causes the cells to discharge, resulting in a chain reaction of neuronal interactions.

Because of its relatively noninvasive nature, TMS is generally considered to be a low-risk technology. However, there are risks associated with TMS. Among the most troubling is the potential induction of seizures. Seizure activity may occur when the induced neuronal excitability spreads beyond the site of stimulation, and it typically involves involuntary hand or arm movements. Although safety studies in healthy, normal participants show few if any side effects, TMS may pose greater risks in individuals suffering from conditions such as Parkinson's disease or major depressive disorder than in those with healthy brain tissue.

The most important risk factor for seizures is the overall health of the person's brain. From 1985 to 1995, seven cases of unintentionally induced seizures occurred during clinical research on repetitive TMS. In 1996, the International Workshop on the Safety of Repetitive Transcranial Magnetic Stimulation reviewed these cases and developed safety guidelines. Since the guidelines were introduced, in 1998, the risks associated with the therapeutic use of TMS have diminished considerably. Even so, the risks of nontherapeutic use remain substantial. Absent appropriate screening to evaluate the state of an individual's brain before TMS is performed, there is a possibility of serious side effects, including seizures.

Since the 1980s, deep brain stimulation has been employed to treat Parkinson's disease and other movement disorders. DBS uses electrical stimulation of specific regions of the brain to control a variety of pathologies. The technology involves the implantation of an electrode into the appropriate region of the brain. A high-frequency electrical current–producing pulse generator is connected to the electrode via an insulated cable called the extension. The extension runs under the surface of the skin to the upper chest cavity, where the pulse generator, a small, battery-controlled, box-shaped device, is implanted. Pulses are typically high frequency (over 100 Hz) and are generated in an on-and-off manner, so that short bursts of stimuli are interspersed with periods of no stimulation.

The mechanism of action of DBS is still not generally understood, but the procedure is generally safe and is supported by scientific research and clinical studies. Current thought is that DBS electrical signals could activate or block neuronal activation or could alter neural networks. DBS has been approved for the treatment of Parkinson's disease and dystonia, and the technology is being tested as a last resort for depression as well as chronic pain, Tourette's syndrome, Alzheimer's disease, epilepsy, and obsessive-compulsive disorder.

As we've seen, there's no grand theory of the brain, at least not yet, but over the millennia the tools have become vastly better. Even so, how well the modern tools work and whether they are safe remains an open question in many cases. Transcranial direct current stimulation (tDCS) is a very simple, noninvasive technology that uses a nine-volt battery and at least two electrodes on the scalp to change electrical activity in neurons, turning them on or off. It's the descendant of ancient brain stimulation methods, such as using electric eels to treat epilepsy. Some hope that tDCS can offer a cheap alternative to more expensive and risky ways of changing brain states, at least for a short time. Beyond therapy for psychiatric and other maladies, tDCS is being investigated to determine whether it can be improved to enhance certain cognitive abilities. Fluid intelligence is the name given to the ability to respond adequately to novel situations. Using specially designed electrodes, the Beckman Institute at the University of Illinois is trying to develop a high-definition form of tDCS to enhance fluid intelligence.

In 2008, the FDA granted Neuronetics approval to manufacture a TMS device for use with treatment-resistant major depressive disorder in adult patients. Although the FDA has not yet approved TMS for other medical purposes, several academic institutions are using the technique experimentally in clinical research settings for the off-label treatment of other brain disorders. For example, researchers at Columbia University have studied the effect of TMS on the memory of students after an extended period of sleep deprivation. Some evidence suggests that TMS also could be employed clinically to suppress traumatic memories. According to a 2004 U.S. Army mental health survey of troops who had fought in the Iraq War, about one in eight reported symptoms of post-traumatic stress disorder, but fewer than half of the affected individuals had sought treatment. It is possible that TMS could provide an effective therapy for post-traumatic stress disorder by helping to suppress traumatic memories and the negative emotions associated with them, or by preventing memory formation in the first place. If TMS turns out to be useful for memory suppression, however, there is a risk that soldiers could be returned to combat too soon after suffering psychological trauma. Such treatments also may have unintended long-term consequences that are not immediately apparent.

It makes sense that the high-definition tDCS project is funded by the Intelligence Advanced Research Projects Activity, under the Office of the Director of National Intelligence. For intelligence operatives, the ability to respond well to new situations is a critical skill. During World War I, IQ tests were used to try to identify the best candidates for clandestine operations. In World War II, personality inventories and situation tests were used. But the results weren't always satisfactory. In the world of spy craft, every generation seems to try whatever new tools are available.

A few studies have explored the use of fMRI and TMS together for the purpose of lie detection, as an alternative to polygraph use. Scientific evidence for the validity of polygraph data is lacking, and much evidence suggests that, in many screening and investigative situations, the predictive value of the technique is poor. According to a patent application for an fMRI/TMS "deception inhibitor," an fMRI scan would first indicate whether or not an individual was attempting to deceive the interrogator, after which TMS would be used to block the deception by inhibiting the relevant part of the cerebral cortex.

BRAIN-MACHINE INTERFACES

No one needs new brain therapies more than people who have lost the ability to move their limbs. In the late 1990s, scientists demonstrated neurological control of the movement of a simple device in rats and, soon thereafter, of a robotic arm in monkeys. More recently, a pilot study of BrainGate technology, an intracortical microelectrode array implanted in human participants, confirmed one thousand days of continuous, successful neurological control of a mouse cursor.

Noninvasive technologies for harnessing brain activity also show promise for human use. Progress recently has been reported on a "dry" EEG cap that does not require a gel to obtain sufficient data from the brain. The brain cap is reported to reconstruct movements of humans' ankle, knee, and hip joints during treadmill walking in order to aid rehabilitation. Researchers have taught monkeys to neurologically control a walking robot on the other side of the world by means of electrochemical measurements of motor cortical activity. It doesn't take any imagination to appreciate how that kind of control could be adapted to military uses. Instead of moving a joystick and pushing buttons to run a drone thousands of miles away—physical actions that often delay crucial instructions from the operator to the device—a brain–machine interface could send instructions at a speed only limited by that of electrons.

Brain–computer interfaces convert neural activity into input for technological mechanisms, from communication devices to prosthetics. An emerging interface system called intracortical microstimulation is a neurologically controlled prosthetic that could send tactile information back to the brain in nearly real time, essentially creating a "brain–machine–brain interface." In addition to devising prosthetics that can supply sensory information to the brain, brain–machine–brain interfaces may directly modify neurological activity. Again, the military applications are provocative. Portable technologies like near-infrared spectroscopy, for example, could detect deficiencies in a warfighter's neurological processes and feed that information into a device utilizing in-helmet or in-vehicle transcranial magnetic stimulation to suppress or enhance individual brain functions, such as those having to do with stress or concentration. We will have more to say about military applications in chapter 8.

The advances in brain–machine interface experiments have come so quickly since the early 2000s that they now are almost passé. How some of these sophisticated laboratory technologies will become part of our ordinary lives is impossible to say, but for some years the simplest versions of nervous system implants have been adopted by a scattered, experimental, and—some might say—eccentric few.

The British engineering professor Kevin Warwick lays fair claim to being among the first human cyborgs.[6] In the late 1990s, Warwick had a radio-frequency identification tag (a relatively simple device that transfers data through an electromagnetic field) implanted in his skin to control lights, heaters, and other computer-controlled devices nearby.[7] Warwick achieved a long-distance version of this feat in 2002, when his implant sent a signal over the internet from the University of Reading in England to achieve a degree of control of a robot arm at Columbia University in New York.[8] A pioneer in his way, Warwick was quickly labeled "Doctor Cyborg" by the press. In a more romantic gesture than Warwick's other experiments, his wife had an electronic microarray implanted in her arm that received a signal from Warwick when he intended it, perhaps paving the way for telepathic communication.[9]

In the past couple of decades, the idea of brain-to-brain interface has moved well beyond those early experiments. Attempts are being made to develop systems that permit communication between brains without language or the other evolutionary-based signals, such as facial expressions or eye movements, that are often subtle (and unconscious) but quite effective. A highly publicized experiment at the University of Washington in Seattle involved a "sender" hooked up to an EEG device that records brain activity and then sends electrical signals over the internet to a "receiver" with a TMS coil near the brain area responsible for hand movements.[10] The receiver's hand moves to do the actual firing at the target when the sender thinks about firing a weapon as part of a computer game. These experiments usually work if the sender is thinking clearly about the *Fire!* command. But since they don't always work, some training is necessary, and even then, the maneuver apparently fails at least 10 to 20 percent of the time.

Still, the brain-to-brain interface experiment is proof that this concept can work in the laboratory, and it suggests that it could be adapted to rehabilitation for people with neurological disorders. Instead of, or along with, traditional psychotherapy, perhaps a future "neurotherapist" could tap directly into a dysfunctional region of an injured person's brain and help heal it by transmitting electrical signals that reestablish a pathway or build a new one. Even if the technologies were someday to become available, however, a goal like that would require years of experiments with very impaired patients under highly controlled conditions. In that scenario, the neurotherapist would have a lot more power over the patient than the traditional psychotherapist. As new neural pathways are being laid down, what rules would govern the goals of the therapy? New ethical standards would have to be crafted by the practitioners to protect the individuality and uniqueness of the patient as never before.

PSYCHEDELICS

Now being rediscovered as potentially respectable options, psychedelics hold promise to alleviate the miseries of anxiety and depression. There was a vigorous period of exploration of psilocybin and LSD from the mid-1950s to the mid-1960s, with about a thousand published papers, until these compounds were gradually classified as being without scientific or therapeutic merit. The Food and Drug Administration, newly authorized to address drug efficacy as well as safety, could not conclude that there was sufficiently rigorous data for approval. In addition, law enforcement authorities and politicians came under pressure due to claims about risks (though one of President Nixon's aides later said this policy was largely political).

As was the case for insulin and diabetes, these advances will be empirical first. Then mechanisms will need to be explored, in the hope of further improvements. A few researchers are being permitted to conduct well-controlled clinical studies on the use of LSD for anxiety caused by life-threatening illness, as well as experiments with MDMA, marijuana, and ayahuasca, a psychedelic tea brewed from plants found in South America. Groups like the Multidisciplinary Association for Psychedelic Studies have organized carefully controlled trials at major medical centers, such as one which used psilocybin to address death anxiety. Psilocybin also is FDA-approved for clinical trials on treatment-resistant depression.

Modern antidepressants called selective serotonin reuptake inhibitors also can increase the availability of serotonin in the brain. The mechanism of action of psychedelics is not well understood, but some evidence points to stimulation of serotonin receptors in the frontal cortex. Neuroimaging has shown that people on psilocybin have decreased activity in the anterior cingulate cortex/medial prefrontal cortex and decreased positive coupling between the medial prefrontal cortex and posterior cingulate cortex. Similar effects have been found in people on LSD, which has helped researchers to understand its hallucinatory quality, as it expands the connectivity of the primary visual cortex. By contrast, the parahippocampal region and retrosplenial cortex showed reduced connectivity, which correlated with less of a sense of ego and more of a feeling of "oneness" with the universe.

Oneness is a fine goal, but the world is still a dangerous place. We live in revolutionary times; the technology is mind-boggling, and the speed of the tools and their applications, as well as their potential for destruction, are formidable. The technology coevolves with neuroscientific discovery. Fusions of brain to brain or of brain to machine, and the growth of tissue within machines, are perhaps within a grasp or two. Although they are often neglected, compared to the more obvious medical needs of warfighters, mental health concerns need to go hand-in-glove with our national defense needs. Both are rooted in the fact that the world poses constant but unpredictable risks, what philosophers such as John Dewey, musicians such as John Cage, and painters such as Marcel Duchamp called the "aleatory" nature of the universe. For us as creatures over evolutionary time, the predators are different but the vigilance is still necessary, and in geopolitical terms, the arms race is a sad reminder. The threat of neural hijacking is not absurd but is merely a more modern form of brain seduction and mind control.

We have come a long way from the horns of Delgado's bull. We know more about the brain and the technology for drug discovery and delivery, and we have some understanding of the chemical milieu. Gene regulation offers far more precision than our committed neuroscientist in the bullring could have imagined. We now are in the midst of building precision medicine with predictive coherence and suitability. The going is still slow, considering the pressing needs of those with neuropsychiatric disease, but it is not a fantasy. Along with the grinding research that is asking questions about what works and what does not, the tools are present and expanding.

8

SOCIALIZING

Prosocial behaviors have to do with helping others. They are socially constructive actions that respond to our desires for reciprocity and fairness. They might be altruistic or just motivated by a sense of goodwill, of human solidarity. Darwin understood prosocial behaviors as the basis for cultural evolution. They facilitate basic social contact through social sentiments of approach or of social avoidance or aversion. The prosocial brain is the biological basis of human values.

A SOCIAL SPECIES

As Aristotle, Darwin, and Dewey observed, we are a social species. Our brains reflect this palpable fact. Therefore, in this chapter we emphasize the prosocial brain, providing a naturalistic background for the origins of morality, which is a diverse appraisal system orchestrated by the brain in suitable environments. The moral decisions that have to be made in doing neuroscience research exemplify the way in which human beings apply their prosocial brains to ethics. Adjudication of ethical decisions in neuroscience takes place among small groups of individuals debating, cajoling, and persuading about whether experiments are worthwhile and whether they cross ethical boundaries. To nonscientists, the process of science might seem strange, highly technical, and laden with jargon, but it's really the same as any group of people getting together to decide what is ethically acceptable and desirable. In this way, the process of neuroscience goes to the heart of human socialization and the ethics of our interactions with others, even while it studies the brain as the biological basis for those interactions.

In philosophy, naturalists believe that values emerge from the conditions encountered in experience, that they are not separate from facts. Nonmoral values crop up, for example, when we realize the importance of something that we previously took for granted, as with our growing awareness of the gradual pollution of drinking water. Moral values often come to the fore in a similar way when we are faced with ethical decisions that we didn't think about before, perhaps because we are facing an unexpected dilemma such as deciding about the care of a loved one near the end of life. At those times, we may find out what's really important to us.

Darwin understood that human moral ideas grow out of our prosocial brain because our evolution is tied to other human beings. This is, in part, why consensus in morals and in science is essential. The democratic view of a community of inquirers has been embraced in science, where conflicting views are not denied or discarded but agreement is sought through the accumulation of evidence and the force of reasoned argument. Despite the ideals of this system, there are places where scientific consensus has not been achieved.

There can be many reasons for a failure of scientific consensus, including a lack of evidence, but then there is always reasonable hope that someday the evidence will be obtained. However, when moral consensus has not been achieved, as in the case of abortion or animal rights, it is often because of basic ambiguities about certain key terms, such as *person*, combined with uncertainties about the extent to which these beings are capable of being self-aware. Our problem-solving capabilities are laden with valuation and appraisal. Ethical appraisals are an essential part of these capabilities and emerged as part of our basic brain activity, but when information is lacking, our appraisals may be stymied.

Human evolution, like our cultural development, is marked by many neural/cognitive events, but social capabilities embrace most of them. One early social behavior involves observing what others are looking at and what they are doing about it. In our social evolution, then, the social brain is tied to daily survival in primate groups. Thus, social behavior is a premium in our evolution, and when this behavior goes wrong, the price can be high. Our species is prone to malignant aggression and devastating destruction. We are wired for both cooperation and conflict. Both reflect expanded cortical tissues. Those more prone to ameliorate conflict may have greater oxytocin expression. Bonobos may provide some evidence for this effect. Compare this to the production of vasopressin, which is more likely to be expressed in the much more violent chimpanzee, the bonobo's full-fledged cousin.

Social problems are pervasive in primate cognition, and a wide array of brain regions, including regions of the temporal lobe, track and perform diverse aspects of social cognition. Also fundamental in our cognitive architecture is whether we perceive an object as animate, and therefore intentional, or not. Different brain regions respond to animate versus inanimate objects; regions of the lateral fusiform gyrus, and the amygdala, are readily activated by animate objects, whereas tool use is linked to Broca's region and other cognitive motor regions. Even very young children are quick to discern intentional events and the possibility of deception and intention. Brain appraisal systems tied to social contact underlie this capability, and diverse regions of the brain are involved in the process. Understanding others' intentions is essential for group formation.

"YUCK!"

Aversion is the flip side of cohesion. Darwin described facial and bodily expressions linked to disgust. Disgust reactions allow organisms to identify objects that don't settle well, like rancid meat, and humans use this information to inform others about what they like or do not. Their origins may lie in basic revulsion to rotting food and the visceral sickness that is one common result. People eventually come to be repulsive or attractive as well.

Revulsion to what we find morally repugnant is a primary feature of approach and avoidance behaviors. The neural systems that traverse the brain from forebrain to brain stem, including primary taste/visceral cranial nerves, underlie this revulsion. As with most ways in which we explore the world, objects are evaluated in an hedonic context, with our evaluation varying depending upon the related sensory system—whether gustatory, olfactory, auditory, or visual—or relevant nonsensory system, such as an appraisal of symmetry, structure, or organization. The gustatory system is a vital part of the limbic brain. It begins in the mouth and ends in the anus and is intimately involved in determining what is in the world and how things affect the gut. Gustatory sensibility impacts moral sensibility.

Disgust also plays a central role in social aversion. Prototypical forms of disgust, associated with nonsocial functions, can be found in nonhumans. Distaste, nausea, and vomiting occur following exposure to potentially toxic or contaminated foods. Odors have a clear adaptive function. In humans, disgust and its close relative, contempt, play a clear role in interpersonal settings as well as in these more primitive contexts. In contrast to anger, disgust and contempt are slower to fade out; they tend to "stick" or to become a property of the disgusting object, intensely devaluing it. Thus, in the same way that neural systems underlying primitive forms of pleasure and social bonding operate in highly complex social situations associated with human cooperation, neural systems underlying aversive responses are related to these bonds. Unfortunately, whole groups of human beings can be stigmatized in this way. Thus, the prosocial can become antisocial and even cruel.

But these vices, too, are part of this prosocial orientation, which includes guilt, embarrassment, compassion, and gratitude, which in turn promote cooperation, helping, reciprocity, reparative actions, and social conformity. A subclass of these, the so-called empathic moral sentiments (guilt, gratitude, and compassion), putatively share the attachment component and play a central role in behaviors linked to empathy. Paradoxically and inevitably, despite their sometime association with inhumane attitudes, prosocial behaviors are at the heart of morality. These behaviors determine what we do for each other, what we owe to each other, and how we get along with each other.

But grounding morality in evolution is not the same as reducing morality to evolution. Rather, it is a marker of the continuity of both our biological and our cultural evolution. Initial disgust reactions that unfortunately manifest themselves in the worst forms of mindless hatred, tribalism, and socially organized cruelties, such as systematic racism, are rooted in ancient systems like the olfactory. Yet the human brain has also developed systems to critically assess and resist what has been called the "yuck" factor, leading to actions such as the elimination of anti-miscegenation laws.

Unfortunately, not all of us are capable of executing prosocial capacities. Neuroscience and moral judgment are rooted in the frontal cortex. Individuals with prefrontal damage are vulnerable to autonomic diminution, and such diminution is thought to increase the likelihood of sociopathic behavior. There can be other impacts of frontal lobe damage, such as decreased activation of skin conduction and visceral responses to facial and bodily expressions. This region is a large part of the brain, and the specifics of how damage and injury impact behavior is not fully understood. Interest in this issue has been reawakened recently by systematic studies of acquired personality changes due to brain damage, mostly to the frontal lobes, and these studies may give us more information on changes resulting from problems in this region and on healthy functioning.[1]

Prosocial values such as friendship and loyalty, and the dispositions they evoke in the agent, could emerge by connecting culturally shaped information, represented in cortical structures (for instance, conceptual and action knowledge related to "friendliness" or "friendly manners," represented in the frontal and temporal association cortexes), with affiliative motivations arising from limbic circuits. This interaction of affiliative experience with social concepts, social perceptual features, and prediction of action outcomes thus provides the basis for sentiments such as compassion, guilt, gratitude, and empathy.

Using fMRI, Joshua Greene and colleagues probed another important aspect of morality.[2] Normal participants were exposed to moral and nonmoral dilemmas that imposed a high load of reasoning and conflict. Moral dilemmas were divided into the moral-personal (in which the agent directly inflicts an injury to another person to avoid a worse disaster) and the moral-impersonal (in which the agent does this in indirect ways, such as by pressing a button, in sacrificing one person to save five). This is the famous trolley dilemma: pushing one person in front of a train to save multiple lives.

In one experimental context, the dilemma is up close and personal, while in another it is more removed. The dilemma, not surprisingly, is more difficult for participants who are up close and personal in this experimental context, indicating that contact is an important aspect of this decision-making. The brain regions measured in the experiment included the frontal gyrus, posterior cingulate cortex, and regions of the temporal and parietal lobe, in addition to the amygdala. These same regions are active under other conditions, so it is not a signature terribly significant to morality, but it is relevant to a larger class of social contexts. This experiment does, however, illustrate that ordinary contact matters. This means that the stuff of embodied experiences, the consideration of others, and the impact of others' experience upon us all are vital to our moral decision-making.

Other studies have addressed additional key issues in moral judgments, including the contribution of general emotional arousal, the presence of bodily harm, response times, semantic content, cognitive load, conflict, intention, consequences versus means, emotional regulation, and justice-based versus care-based judgments.[3] One additional study from Greene et al. showed that the greater the cognitive load, the more inference with utilitarian judgments.[4] This indicates that time, the complexity of context, and the demand of cognitive control impact judgment.

Studies like these have extended our knowledge of the neural substrates of moral judgment and emotions. However, despite these advances, which were made possible in large part due to a substantial development in functional imaging techniques, the roles of subcortical/limbic structures in morality remain obscure. This can be explained both by technical limitations (fMRI is intrinsically less sensitive to detecting activity in those regions) and by the lack of robust models and detailed knowledge about the role of specific subcortical/limbic circuits and their functional relationships with cortical regions in the context of human moral cognition. The greater the conflict, and perhaps the more scenarios to consider in the so-called appropriate condition, the longer the reaction time.

Philosophers have long wondered how the moral personality develops. It is widely assumed that the brain isn't on its own irrevocable course from birth. Aristotle emphasized that, when we are young, our values are shaped by those with whom we choose to associate, so we should choose wisely (and he did believe that we were free to choose). If that's true, then the brain itself must be modifiable. At the end of the nineteenth century, the philosopher and psychologist William James contributed the concept of neuroplasticity, noting that brain circuits must be "weak enough to yield to an influence but strong enough not to yield all at once."[5] But the idea of change in the adult brain remained a matter of hot debate until evidence from various sources, including from the staining of neural tissues, became overwhelming.

One way to capture the plasticity of function is through gene regulation, such as histone modification of RNA and RNA splicing and editing. Up to 40 percent of the genome is linked to transposable elements that are subject to change. This capability underlies structural and functional features that in turn underlie dendritic remodeling of regions for memory and learning.

Remodeling hippocampal tissue is a lifelong process and is essential for continuous learning and memory. Adrenal steroids penetrate the brain and bind to the hippocampus. Glucocorticoid hormones are steroids that are a workhorse of bodily maintenance and stability. Steroids like cortisol have both genomic and nongenomic impact on bodily tissue, including the brain. One impact is to induce gene regulation and gene expression; the faster-acting impact of steroids is on membranes. Both are critical for working memory and learning. There also are a number of intracellular and extracellular signaling systems (including endocannabinoids, excitatory amino acids, and brain-derived neurotrophic factors) that form the basis of these events. Some of them, such as cannabinoids, involve fast-acting, membrane-related events or nongenomic events, and others, like brain-derived neurotrophic factors, involve genomic or more slowly related events.

As Elizabeth Gould, Bruce McEwen, Joe Herbert, and many others have made clear, rebuilding and sustaining neural tissue is a lifelong and continuous series of events across multiple neural regions.[6] Neurogenesis (the growth of new neurons in select regions of the brain, via neural pruning) underlies the expansion and shrinkage of neural tissue. Regions of the amygdala expand under duress, whereas the hippocampus declines, as can occur during bouts of post-traumatic stress disorder or depression. Information molecules, such as the brain-derived neurotrophic factor, also are involved with changes. They underlie everything we do, including our social behaviors.

Impoverished or enriched environments deeply affect the brain. These are not small results. Epigenetic neural changes occur across the life span. Some periods are more critical than others, particularly early in development. An enriched environment reduces lateral amygdala hypertrophy, which can occur in stressful and adverse environments correlated with elevated cortisol. We have known this for some time, but neuroscience just gave another edge to our education by informing us of the debilitating impacts of poverty and other environmental factors on the brain. This knowledge can inform our understanding of how these effects can shape priorities in public policy, like funding for early childhood education. Positive forms of social contact, including adult support, can ameliorate cortisol levels.

GROUP FORMATION

Coalitions and group formation are fundamental in human life. We form groups easily, and with good reason, because in general primate groups enjoy greater survival rates than do isolated individuals. Social capability is tied to infant and adult survival. Because we humans undergo a prolonged juvenile stage to enable a foothold in the world, human social ties can be complex and convoluted. They involve interactions across hierarchical levels and degrees of both psychological and physical closeness. In other words, social capability is tied to infant and adult survival. When one compares young children and related primates, solutions to physical problems are similar, early in development, but it is very different for social problem solving, which is where a lot of human development is concentrated. Size, complexity, and social contact, in other words, are consistently linked to neocortical expansion. The larger the coalitions of both carnivores and primates, the greater the cortical expansion. This holds for most of the neocortical regions of the brain.

The field of social network analysis long ago developed sociograms, social graphics that "map" human relations, defined in terms of interpersonal selection and rejection, based on any number of criteria. Once understood, those patterns can be used for any number of purposes, from finding ways to better integrate isolated individuals into social life to identifying "opinion leaders" who will help sell products to running propagandistic influence campaigns. Where sociometrists once were limited to laboriously hand-drawn group maps of relatively small numbers of individuals, those networks are now routinely produced by social media platforms, often in real time, over countless criteria of interpersonal choice.

Conferences of neuroscientists are no different from any human system. Like gatherings of any large number of people, navigating a meeting of thousands of attendees is tied to diverse social engagements, coalitions, and competition among laboratories— as well as a handy downloadable cell phone app that tells you which speaker is in which session and how to find your way across a mammoth convention center to the restrooms and snack bar. Forming alliances and finding one's way around the gargantuan exhibits hall are nontrivial events in our social engagement. Getting from one end of the convention to another in time for the next paper session is hard. What is even harder is putting the diverse ways in which to understand the brain into a science of how we imagine the brain underlies mental function. That's what science meetings, as well as some nice dinners with colleagues, are for.

No less than any other human event, the neuroscience meetings are a scene of cultural selection where a battle of ideas (often barely distinguishable from a battle of personalities) takes place. Cultural selection is a form of evolution that subsists through all the mingling, networking, glad-handing, exchanging of reprints, speculating about data, drinking, overeating, gossiping, and negotiating of research grants and contracts. In theory, the ideas that have borne intellectual fruit, have been replicated in the field of inquiry, have garnered interest and support, and, perhaps most of all, have provided understanding will make their presence felt in an indefinite series of meetings, large and small, local and international. This battle of ideas in the process of cultural selection may not always be a genteel process. Often, clashes of various sorts underlie the fabric of scientific engagement. These clashes lead to forms of Darwinian selection, determining the science that breaks and the scientists who gain favor. Notably, though, Darwin (and Peter Kropotkin) emphasized that cooperation is a fundamental feature in our evolution, despite the presence of deceit, bias, falsification, and endless human limitations on human reasoning.[7] Cooperation is as critical as competition in science, because we need to learn from one another and to develop new ideas. Enter most neuroscience labs and you will see the younger workers learning from the older ones, bearing a not wholly accidental resemblance to a clan of nitpicking orangutans. Learning, making gains, and elaborating on old ideas entails cooperation and engagement with one another.

Young children and our closely related primate cousins develop solutions to physical problems in similar ways, but they solve social problems very differently, and this is where much of human development is concentrated. Increased cerebral tissue, connections between sites, and the production of neuropeptides and neurotransmitters make up cortical expansion. The range of cerebral tissue mass reflects the capacity to compute and thereby thrive in the social milieu. Size, complexity, and social contact, in other words, are consistently associated with neocortical expansion.

Group size affiliation also has been consistently linked to cerebral expansion across primates (simians, hominoids, humans), extinct or not. In both carnivores and primates, the larger the coalitions, the greater the cortical expansion. This principle holds for most of the neocortical regions of the brain. This joint social capability is linked to brain region expression and expansion. Innovative tool use is likewise linked to cortical expansion. Not surprisingly, neocortical expansion is also associated with deceptive behaviors. Deception is a feature of primate behaviors generally, but so are trust, reliability, predictive coherence, and expected outcomes.

Keeping track of who does what in the group is one of the daily cognitive chores that underlie social interaction. In our social evolution, the prosocial brain is tied to daily survival in primate groups. Language, tool use, motor dexterity, binocular vision, and social grooming are among the many results of the socially evolving brain. For most of us, dorsal and ventral neural visual fields of information processing permeate the location and meaning of events, tied to the organization of action.

The human neocortex is much larger than that of our closest relatives, the bonobo and chimpanzee, and our evolution is tightly linked to prosocial capabilities. The anti-adaptive results of aberrations in social function, such as autism, reveal the importance of this core function. As group size has increased in hominoid evolution, cortical regions have expanded. The lateral region of the amygdala, more closely tied to neocortical evolution and function, also has expanded in size as it is related to social complexity.

Scientific hypothesis formation might seem like deadly serious stuff, but underneath it all, it is a form of play. Play behaviors aren't just for fun. Play with ideas, the drudgery of test and failure, the excitement when something works, and, even more importantly, reliable replication are all common themes, even in children's play. Infant survival in a number of closely related species is tied to the sociality that comes with play. At least some of the popular fascination with the genius of Albert Einstein has to do with his ability to engage in "thought experiments," mental play that helped him achieve profound insights. Brilliant though Einstein was, his individual play alone would have been useless if it hadn't been accessible others, so, as Einstein played with ideas, he also elaborated the rules that enabled others to play in his mental sandbox. In the end, even Einstein's brain was only human and deeply prosocial.

Social contact is also one form of the self-regulation of neural/behavioral activation. We reach out to what we need, want, would like to know about, and need help with, all of which activates the brain's self-regulation. Social contact is essential for our internal state. A hyperactive amygdala can be ameliorated by reassuring social contact. The regulation of the neural milieu is likely due to the safety, assurance, and predictive reliability provided by consistent social contact. The social dimension of primate evolution directly affects diverse information molecules. Indeed, the "social friendships" that are exhibited in primate grooming behavior have been shown to reduce potentially dangerous cortisol levels associated with stress. Social stability and social prediction are therefore keys to the regulation of many brain regions, from cortical to subcortical. The brain invests a lot of neurons in navigating the social milieu, evidence that being social is fundamental for survival.

From stacking plastic blocks to building a bridge to postulating a scientific theory, problem solving is driven by appetites. When the solution is at hand, we enjoy what John Dewey called a consummatory experience. Information molecules in the brain underlie the appetitive and consummatory experiences associated with diverse social and problem-solving behaviors regulated by the brain. Consummatory experiences are events that are tied to the satisfaction of curiosity, craving, or desire. The "search mode" in our experiences is tied to satisfying curiosity, to the need to solve a problem. The satisfaction is tied to the resolution of the problematic.

Information molecules such as vasopressin and dedicated neuronal circuits underlie such events and constitute part of the neuroscience of search and discovery. They are part of the brain's social signaling systems, like the neuropeptides that underlie social contact. Other signaling systems include neurotransmitters. One form of the serotonin receptor has been consistently linked to social gregariousness or social approach. Behaviorally withdrawn, temperamentally shy children have increases in amygdala activity in response to novel events, reflected in differences in the short version of the serotonin and dopamine receptor subtypes. Alterations in gene expression for the neurotransmitter serotonin also are consistently linked to variants of antisocial and retaliatory behavior.

Problem solving is a social activity. The wholly isolated inventor or creator is a myth. Observing and learning from others is critical for a great many mammals, not only human beings. Omnivorous rats watch and learn from the eating habits of other rats, for instance. Not surprisingly, neurons in diverse regions of the brain can mirror the movements and the organization of members of the same species. This basic arrangement enables a foothold in a world of pedagogy and lifelong learning by enabling an individual to replicate behaviors that it sees in others. The anterior cingulate cortex is linked to basic approach and avoidance and the decision-making necessary for such events.

Language is another key feature of human socialization. The advent of language, including the simple but crucial ability to share a joke or gossip, core functions made possible by our prosocial brain, changed everything for our species. Nearly every dimension of our lives is dominated by language, and it is a key tool in our development and advancement, allowing individuals to coevolve in a mosaic of capabilities. By linking with others in small groups, humans have been able to make discoveries and share ideas across time, leading to what we now describe as culture—including that particular form of culture known as science. There are diverse functional links between language use and motor expression, between syntax and diverse regions of the brain that underlie syntactical ordering of events.

SOCIAL DEVOLUTION

Autism is a disorder that reflects devolution of function. It involves withdrawal from social contact, inability to discern others' general interest, and problems making effective eye contact. By contrast, people with Williams syndrome, a genetic deletion, are exceptionally outgoing and extremely socially engaged. They also have intellectual disabilities in areas that are essential for normal functioning and tend to exhibit an overall decrease in cortical volume in the outermost layer of the brain, which is tied to facial expressions and decreased inhibition. Alterations in both oxytocin and vasopressin, two information molecules tied to social regulation, are also found in Williams syndrome; prosocial information molecules such as oxytocin tend to be higher in those with Williams syndrome and lower in autistic people. Thus, one condition is an exaggeration of prosocial behaviors and the other is devolution of social contact. Core differences between the two groups reflect the phenotypic features of their brains: cerebral volume is actually larger in autism, but size does not add up to what one expects. However, regions of the temporal cortex are larger, in general, in Williams patients. The cerebellum is normal and the brain stem is decreased, when compared to normal controls.

Williams syndrome shows that social capabilities exist within the context of other forms of cognitive/neural function that make what we do possible. Individuals with Williams syndrome tend to have a lower IQ than is typical in the general population, although there is a fair degree of variance. Their spatial capabilities are quite compromised, but their linguistic capabilities can be excellent. Regions of the temporal lobe, or what is often called the ventral stream, essential for encoding social significance, are intact. The more dorsal stream, associated with the regions of the temporal lobe of the cortex, including the amygdala, encodes the organization of action and enables assessment of the meaning of events.

Twenty to thirty genes or sets of specific genes are selected for prosocial behaviors; an aberration in one of these leads to Williams syndrome. Sets of specific genes (such as GTF21) impact the volume of the insular cortex and connectivity to regions of the brain, including the amygdala. Other genes (such as LIMK1 and CL1P2), essential for neuronal development and maturation, are tied to Williams syndrome, as is chromosome 7q11.23 hemideletion.

Interestingly, eye contact is more apparent in females than in males early on in ontogeny, and more males than females tend to manifest as autistic. Eye contact is a threat indication in most primates. But, for humans, shared contact is something that evolved with us, and eye contact is less directly connected with frankly aggressive behavior, although it can be perceived as a threat in the right circumstances. At the same time, we are just as likely to distrust someone who can't look us in the eye.

AND FINALLY

This chapter has highlighted a theme we have tried to convey throughout *The Brain in Context*: the dominance of the social milieu. Our individual well-being is inextricably linked to others.

In this social milieu, human meaning and contact is a measure of our well-being and our social solidarity. Even being alone, our ability to be immersed in reverie during solitary walks, requires a social baseline, as beautifully described by Jean-Jacques Rousseau.[8] Human meaning lies in the diverse forms of social contact, and it is lifelong.

And the social brain is the great bulk of the brain. It is all of the cortex, old and new, a never-ending encoding to entail capability and adjudication as we navigate the social spaces, as we step into the moral fabric and conflict-ridden spaces of conflict, amid reprieves of relief. Our lives are still nasty, as they were in Hobbes's time, but often less brutish and short. Indeed, they are far longer than what Hobbes could have fathomed, and much better for many of us, though surely not for all. Some, like Steven Pinker, emphasize the progress, while others, like Niall Ferguson, remind us of the risks.[9]

We started this book claiming that we would avoid reducing the brain to one characterization or another. Yet we have ended with a chapter on the prosocial brain. To us, this was unavoidable. Our social proclivities set the stage, along with the culture, for the moral appraisals and behavior that we display, cultivate, and value. There may be no special circuit in the brain for moral appraisals or behavior, but prosocial behaviors are a feature of the working of our species. Although truth telling competes with a number of other motivations in our brain, our prosocial sensibilities underlie the fact that we nonetheless usually come pretty close to sincerity. Correcting for our less than wholly truthful tendencies underlies the communal process we call science.

Scientists, like everyone else, are bound to social groups—first small, when they are children, and then larger, when they run their labs, attend department meetings, and schmooze with colleagues. We are relatively big-brained, and our brain expands to our guts. Both are involved in prosocial sensibility. And scientific inquiry takes place within this context.

The historian of science Steven Shapin has outlined a social history of truth gathering and the origins of experimental orientation found in Robert Boyle.[10] Our social nature runs through neuroscience as it does through everything else about us. Another feature of neuroscience is less about theory than about technology and the scientific experimentation of possibilities. This chapter has emphasized our social proclivity as a species, to provide a mirror into the larger social milieu. Social adaptability and capability are at the heart of our evolution.

Our brain is rooted in everyday transactions with one another. Not surprisingly, a great deal of our brain is tied to social knowledge or memory and capability, in one form or another. As historians of science, particularly Thomas Kuhn in his later work, have argued for some time, biology and culture meet at every conceptual and practical space we inhabit as a species. Nowhere is this more manifest than in our social milieu, the transactions we have with one another. The study of the brain reveals just that.

The brain's social space is filled with tools of investigation, of inquiry, of discovery. These tools enable us to dissect the genes that underlie the production of information molecules and their epigenetic or plasticity in neural tissue. This is mostly profoundly expressed in our social needs, manifested in self-sacrifice and self-protection. Although we may think of ourselves as individuals, the truth is that we are designed to work together, revealing our evolutionary drive toward social cooperation and our neurodevelopmental proclivity toward shared decision-making. Darwin and Dewey emphasized the large social space in which our moral landscape is woven. Our success as a species is the extent to which we encourage the social milieu tied to pedagogy, social and ecological development, and human dignity. That species-wide endeavor was always vulnerable. But now it is also placed within considerations of the study of the brain—our brain, those of related species, and those of distantly related species that co-inhabit our planet.

The neural apparatus designed to foster social cohesion in small groups has been expanded to larger groups. We are tied to others, anchored to others. Investigation into the brain, into what we allow or not, agree to or not, is a window into ourselves. Importantly, neuroscience investigations reach out across national borders, to the larger community of inquirers. As we investigate our own neural function, we are tied to humanity as a whole and to our phylogenetic past.

No one is an island, not even a neuroscientist.

NOT THE LAST WORD

T his book has highlighted just a fragment of the rich array of discoveries and problems in neuroscience. We have tried to give a view that is balanced with both excitement and caution. We are excited about this age of neuroscience in which we find ourselves, yet we don't want to understate the difficult road ahead, in terms both of the challenges of gaining new knowledge about the brain and of the ethical issues that will be encountered along the way. Moreover, we have wanted to place neuroscientific considerations in the context of the whole brain, especially in the context of our evolution, both biological and cultural.

Ralph Adolphs outlined some of the outstanding questions in neuroscience:

1. Problems that are being solved, such as neural design and function or computational expression.
2. Problems that we should be able to solve in fifty years or so, including circuit organization, in vivo clarity of connectome, and site-specific and neuronal-specific organization whole-brain maps, and brain decision-making.

3. Problems whose solutions are far down the road, including diverse computational systems in the brain and their interaction, and the expression of behavior.

4. Problems we may never solve, such as exactly how the brain computes or the design principles of specific and general capabilities, as well as big questions that may linger without solution.[1]

We have called attention to the fact that the brain is not only in the head but also is expressed throughout the body and through neuronal interactions within its ecological and social milieu. In appraising events, the brain reaches into the periphery—the heart, the gut, and everywhere else. Aristotle mistook the heart as the organ of thought, perhaps because it seems to register reactions to events in such a dramatic way. Then it became identified more strictly with emotionality and divorced from thought. But, even metaphorically, the heart is not on one side and the brain on the other. All sorts of information molecules inhabit and manage the heart and the brain. The vagus nerve sends direct signals to and from the heart in the organization of action and in making sense of the world we are in. The body reaches out as tentacles in understanding the larger environment. Our concept of epineuromics emphasizes the larger environment in which to understand the neural sciences, the supporting environment in which the brain functions and without which it could never have been created.

In the end, the study of the brain is about us and our epistemic endeavors, our struggle to understand, manage, and flourish. In that sense, the experimental study of the brain is special. It is not just any organ. Rather, the intense focus on the brain illustrates how much of its substance is about what makes human beings human. Those of us who believe that the human experience has intrinsic value—for all its finitude, tragedy, and comedy—need to appreciate the brain in context.

ACKNOWLEDGMENTS

J ay would like to thank in memory Eliot Stellar and
George Wolf, wise neuroscientists and mentors, and many
other colleagues, some long gone. He also thanks many
living colleagues—too many to mention—in the neurosciences.
But here are some: Kent Berridge, Harvey Grill, Joe Herbert,
Joe LeDoux, Bruce McEwen, Ralph Norgren, Jeff Rosen,
Larry Swanson, Peter Sterling, and many not noted here. He
feels so fortunate to have participated in neuroscience with them.
Jonathan is grateful in memory to John J. McDermott, who
introduced him to the work of William James, the grandfather of modern neuroscience, and to the other philosophical
mentors of his youth, Peter Caws, Richard S. Rudner, and Eve-
lyn U. Shirk.

For several years, this project has provided us with a reward-
ing dialogue—and a certain local Washington, DC, bakery and
coffeehouse with two regular customers. We are especially grateful to our editor, Eric Schwartz, for his confidence that we could
write a different kind of book about the brain. Each of our
spouses is no doubt grateful that we provided each other with
someone else to talk with about our work, but we are more grateful to them. Our kids, two sets of two, will just have to inherit
another book.

We both claim close connections to the University of Pennsylvania, where Jay was a graduate student and faculty member and Jonathan is a faculty member. This book is dedicated to our many supportive friends and colleagues at Penn.

NOTES

INTRODUCTION

1. Human Brain Project, "Analysing the Brain with Atlases," accessed May 27, 2019, https://www.humanbrainproject.eu/en/explore-the -brain/atlases; Howard Hughes Medical Institute, "Epigenomic Map of the Developing Brain," July 4, 2013, https://www.hhmi.org/news /epigenomic-map-developing-brain.
2. McCulloch (1965).
3. Neurath (1944).

1. ELECTRIFYING

1. Nishimoto et al. (2011).
2. Kosfeld et al. (2005).
3. Toyama et al. (2014).
4. Delgado (1971).
5. Osmundsen (1965).
6. Gross (1998).
7. Rauschecker and Scott (2009).
8. Flynn (1972).
9. Valenstein (1973); Berridge and Valenstein (1991).
10. Valenstein (1973).
11. Valenstein (1973).
12. Berridge (2019); Berridge and Robinson (1998); Berridge et al. (2005).
13. Sanford et al. (2017).
14. Valenstein (1973); Valenstein (2006).

15. Cajal (1906).
16. Golgi (1906); Cajal (1906); Sherrington (1906).
17. Valenstein (2006).
18. Cannon (1915).
19. Miller (1957); Miller (1965).
20. Koob and LeMoal (2005).
21. Farah (2018).

2. CONSTRUCTING

1. Sterling and Laughlin (2015), 51.
2. Gibson et al. (2019).
3. EPFL, "Blue Brain Project," https://www.epfl.ch/research/domains /bluebrain/.
4. Vrselja et al. (2019).
5. Clark (1998).
6. Friston et al. (2017).
7. James (1899).
8. Chalmers (2003).
9. Nagel (1979).
10. Wittgenstein (1953, 1968).
11. McDermott (2007).
12. Dennett (2017).
13. Bostrom (2014).
14. Reynolds (2017).
15. Silver et al. (2017).
16. Tegmark (2017).

3. EVOLVING

1. Dobzhansky (1962).
2. D. E. Lieberman (2011), 317.
3. Herculano-Houzel (2016).
4. Lilly (1967).
5. Hawrylycz et al. (2012); Hawrylycz et al. (2016); Lein et al. (2007).
6. Paabo et al. (2004); Paabo (2015).
7. McClintock (1951).

8. Mithen (1996); Mithen (2006).
9. Jacobowitz (2006).
10. Adolphs et al. (1998).
11. Richter (1952).
12. Maclean (1990).
13. Swanson (2011).
14. Diamond (2001).
15. Schultz et al. (2008).

4. IMAGING

1. Kanwisher et al. (1997); Kanwisher (2006).
2. Phelps et al. (2000).
3. Lein et al. (2007).
4. Satel and Lilienfeld (2013).
5. Langleben et al. (2005).
6. Langleben and Moriarty (2013).
7. Schacter (1996); James (1890).
8. Kosslyn (1984).
9. Garrett et al. (2016).
10. Horikawa (2013).
11. No Lie MRI, http://www.noliemri.com/.
12. Langleben and Moriarty (2013).
13. Greene and Paxton (2009).
14. State of Tennessee v. Idellfonso-Diaz, 2006 WL 3093207 (Tenn.Crim. App. 2006).
15. Vecchiato et al. (2011).
16. Plassmann et al. (2008).
17. NeuroFocus (2011).
18. Neurosense, http://www.neurosense.com/.
19. Hayden, Pearson, and Platt (2011).
20. Nave et al. (2018).

5. ENGINEERING

1. Anderson (2012).
2. James (1890).

3. Defense Advanced Research Projects Agency (DARPA), "Targeted Neuroplasticity Training (TNT)," https://www.darpa.mil/program /targeted-neuroplasticity-training.
4. Goldstein (1939); Davis and Whalen (2001).
5. Morgan, Schulkin, and LeDoux (2003).
6. Chung et al. (2013).

6. SECURING

1. Kamienski (2012).
2. National Research Council (2008).
3. Rohde (2013).
4. Cristakis and Fowler (2007).
5. Falk and Scholz (2018).
6. Chiao et al. (2008).
7. *Nature* (2003).
8. Eaglen (2016).
9. Defense Advanced Research Projects Agency (DARPA), "Bridging the Bio-Electronic Divide," https://www.darpa.mil/about-us/bridging -the-bio-electronic-divide.
10. Emerging Technology from the arXiv (2018).
11. Pais-Vieira et al. (2015).
12. Devlin (2015).
13. Shanahan (2019).
14. Moreno (2012).
15. Kissinger (2018).

7. HEALING

1. Koob (2015).
2. James (1910).
3. Schaffer (2006).
4. Merton and Morton (1980).
5. Barker, Jalinous, and Freeston (1985).
6. Kevin Warwick's website, http://www.kevinwarwick.com/.
7. Connor (1998).
8. Warwick et al. (2003).

9. Warwick et al. (2004).

10. Stocco et al. (2015).

8. SOCIALIZING

1. Damasio (1994); Damasio (1996).

2. Greene et al (2001); Greene (2013).

3. Moll et al. (2005); Moll et al. (2006).

4. Greene et al. (2008).

5. James (1890), 105.

6. Gould et al. (1996); McEwen (2016); Herbert (1993).

7. Darwin (1859); Darwin (1871); Kropotkin (1902).

8. Rousseau (1782, 1992).

9. Pinker (2018); Ferguson (2018).

10. Shapin and Schaffer (2011).

NOT THE LAST WORD

1. Adolphs (2015).

REFERENCES

Abe, N. (2011). "How the Brain Shapes Deception: An Integrated Review of the Literature." *Neuroscientist* 17 (5): 560–574.

Abe, N., and J. D. Greene. (2014). "Response to Anticipated Reward in the Nucleus Accumbens Predicts in an Independent Test of Honesty." *Journal of Neuroscience* 34 (32): 10564–10572.

Adkins-Regan, E. (2005). *Hormones and Animal Social Behavior.* Princeton, NJ: Princeton University Press.

Adolphs, R. (2015). "The Unsolved Problems of Neuroscience." *Trends in Cognitive Science* 19 (4): 1–3.

Adolphs, R., D. Tranel, and A. R. Damasio. 1998. "The Human Amygdala in Social Judgment." *Nature* 393: 470–474.

Akil, H., J. Gordon, R. Hen, J. Javitch, H. Mayberg, B. McEwen, M. J. Meaney, and E. J. Nestler. (2018). "Treatment Resistant Depression: A Multi-Scale, Systems Biology Approach." *Neuroscience and Biobehavioral Reviews* 84: 272–288.

Akil, H., M. E. Martone, and D. C. Van Essen. (2011). "Challenges and Opportunities in Mining Neuroscience Data." *Science* 331 (6018): 708–712.

Albert, F. W., Mehmet Somel, Miguel Carneiro, Ayinuer Aximu-Petri, Michel Halbwax, Olaf Thalmann, Jose A. Blanco-Aguiar, Irina Z. Plyusnina, Lyudmila Trut, Rafael Villafuerte, Nuno Ferrand, Sylvia Kaiser, Per Jensen, and Svante Pääbo. (2012). "A Comparison of Brain Gene Expression Levels in Domesticated and Wild Animals." *PLOS Genetics* 8 (9): 1–15.

Alembert, J. d'. ([1751] 1963). *Preliminary discourse to the Encyclopedia of Diderot*. New York: Library of Liberal Arts.

Alivisatos, A. P., M. Chun, G. M. Church, K. Deisseroth, J. P. Donoghue, R. J. Greenspan, P. L. McEuen, M. L. Roukes, T. J. Sejnowski, P. S. Weiss, and R. Yuste. (2013). "The Brain Activity Map." *Science* 339 (6125): 1284–1285.

Allen, W. E., M. Z. Chen, N. Pichamoorthy, R. H. Tien, M. Pachitariu, L. Luo, and K. Deisseroth. (2019). "Thirst Regulates Motivated Behavior Through Modulation of Brainwide Neural Population Dynamics." *Science* 364 (6437): eaav3932.

Altman, J. (1966). "Autoradiographic and Histological Studies of Postnatal Neurogenesis." *Journal of Comparative Neurology* 124 (4): 431–474.

Alvarez-Buylla, A., M. Theelen, and F. Nottebohm. (1988). "Birth of Projection Neurons in the Higher Vocal Center of the Canary Forebrain Before, During, and After Song Learning." *Proceedings of the National Academy of Sciences of the United States of America* 85 (22): 8722–8726.

Amaral, D. G., J. L. Price, A. Pitkanen, and S. T. Carmichael. ([1992] 2000). "Anatomical Organization of the Primate Amygdaloid Complex." In *The Amygdala: Neurobiological Aspects of Emotion, Memory, and Mental Dysfunction*, ed. John P. Aggleton, 1–66. New York: Wiley-Liss.

Anderson, D. (2012). "Optogenetics, Sex, and Violence in the Brain: Implications for Psychiatry." *Biological Psychiatry* 71 (12): 1081–1089.

Aristotle. ([350 BC] 1968). *De anima*. Trans. D. W. Hamylin. Oxford: Oxford University Press.

Arlotta, P. (2018). "Organoids Required! A New Path to Understanding Human Brain Development and Disease." *Nature Methods* 15: 27–29.

Asfaw, B., T. White, O. Lovejoy, B. Latimer, S. Simpson, and G. Suwa. (1999). "*Australopithecus garhi*: A New Species of Early Hominid from Ethiopia." *Science* 284 (5415): 629–634.

Asok, A., A. Draper, A. F. Hoffman, J. Schulkin, C. R. Lupica, and J. B. Rosen. (2017). "Optogenetic Silencing of a CRF Pathway from the Central Nucleus of the Amygdala to the Bed Nucleus of the Stria Terminalis Disrupts Sustained Fear." *Molecular Psychiatry* 23 (4): 914–922.

Axelrod, R. (1984). *The Evolution of Cooperation*. New York: Basic Books.

Bale, T. L. (2014). "Lifetime Stress Experiences: Transgenerational Epigenetics and Germ Cell Programming." *Dialogues in Clinical Neuroscience* 16 (3): 297–305.

Banting, F. G. (1937). "Early Work on Insulin." *Science* 85 (2217): 594–596.

Baumgartner, T., M. Heinrichs, A. Vonthathen, U. Fischbacher, and E. Fehr. (2008). "Oxytocin Shapes the Neural Circuitry of Trust and Trust Adaptation in Humans." *Neuron* 58 (4): 639–650.

Bard, P. (1939). "Central Nervous Mechanisms for Emotional Behavioral Patterns in Animals." *Research Publications of the Association for Research in Nervous and Mental Disease* 19: 190–218.

Barger, N., L. Stefanacci, and K. Semendeferi. (2007). "A Comparative Volumetric Analysis of the Amygdaloid Complex and Basolateral Division in the Human and Ape Brain." *American Journal of Physical Anthropology* 134 (3): 392–403.

Barker, A. T., R. Jalinous, and I. L. Freeston. (1985). "Non-invasive Magnetic Stimulation of Human Motor Cortex." *Lancet* 325 (8437): 1106-1107.

Baron-Cohen, S. ([1995] 2000). *Mindblindness: An Essay on Autism and Theory of Mind*. Cambridge, MA: MIT Press.

Barrett, L. (2011). *Beyond the Brain: How Body and Environment Shape Animal and Human Minds*. Princeton, NJ: Princeton University Press.

Barrett, L., and P. Henzi. (2005). "The Social Nature of Primate Cognition." *Proceedings of the Royal Society B: Biological Sciences* 272: 1865–1875.

Barrett, L., P. Henzi, and D. Rendall. (2007). "Social Brains, Simple Minds: Does Social Complexity Really Require Cognitive Complexity?" *Philosophical Transactions of the Royal Society of London B: Biological Sciences* 362: 561–575.

Barton, R. A. (1998. "Visual Specialization and Brain Evolution in Primates." *Proceedings of the Royal Society B: Biological Sciences* 265: 1933–1937.

Barton, R. A. (2004). "Binocularity and Brain Evolution in Primates." *Proceedings of the National Academy of Sciences of the United States* 101 (27): 10113–10115.

Beck, A. T. (1967). *Depression: Causes and Treatment*. Philadelphia: University of Pennsylvania Press.

Bechara, A. (2005). "Decision Making, Impulse Control, and Loss of Willpower to Resist Drugs: A Neurocognitive Perspective." *Nature Neuroscience* 8: 1458–1465.

Bechara, A., H. Damasio, A. R. Damasio, and G. P. Lee. (1999). "Differential Contributions of the Human Amygdala and Ventromedial

Prefrontal Cortex to Decision-Making." *Journal of Neuroscience* 19 (13): 5473–5481.

Békésy, G. von. (1959). *Experiments in Hearing*. New York: McGraw-Hill.

Bernaerts, S., J. Prisen, E. Berra, G. Bosmans, J. Steyaert, and K. Alaerts. (2017). "Long-Term Oxytocin Administration Enhances the Experience of Attachment." *Psychoneuroendocrinology* 78: 1–9.

Bernard, C. ([1865] 1957). *An Introduction to the Study of Experimental Medicine*. New York: Dover.

Berns, G. S., S. M. McClure, G. Pagnoni, and P. R. Montague. (2001). Predictability Modulates Human Brain Response to Reward." *Journal of Neuroscience* 21 (8): 2793–2798.

Berridge, K. C. (2019). "Affective Valence in the Brain: Modules or Modes?" *Nature Reviews Neuroscience* 20: 225–234.

Berridge, K. C., J. W. Aldridge, K. R. Houchard, and X. Zhuang. (2005). "Sequential Super-Stereotypy of an Instinctive Fixed Action Pattern in Hyper-Dopaminergic Mutant Mice: A Model of Obsessive Compulsive Disorder and Tourette's." *BMC Biology* 3: 4.

Berridge, K. C., and T. E. Robinson. (1998). "What Is the Role of Dopamine in Reward: Hedonic Impact, Reward Learning or Incentive Salience?" *Brain Research Reviews* 28: 309–369.

Berridge, K. C., and E. S. Valenstein. (1991). "What Psychological Process Mediates Feeding Evoked by Electrical Stimulation of the Lateral Hypothalamus?" *Behavioral Neuroscience* 105 (1): 3-14.

Berthoz, A. (2000). *The Brain's Sense of Movement*. Trans. Giselle Weiss. Cambridge, MA: Harvard University Press.

Berthoz, A., and J.-L. Petit. (2008). *The Physiology and Phenomenology of Action*. Trans. C. Macann. Oxford: Oxford University Press.

Bielsky, I. F., and L. Young. (2004). "Oxytocin, Vasopressin, and Social Recognition in Mammals." *Peptides* 25 (9): 1565–1574.

Bickart, K. C., C. I. Wright, R. J. Dautoff, B. C. Dickerson, and L. F. Barrett. (2011). "Amygdala Volume and Social Network Size in Humans." *Nature Neuroscience* 14 (2): 163–162.

Blakemore, S., and J. Decety. (2001). "From the Perception of Action to the Understanding of Intention." *Neuroscience* 2: 561–567.

Bliss, M.. (1982). *The Discovery of Insulin*. Chicago: University of Chicago Press.

Bonin, G. von, ed. (1960). *Some Papers on the Cerebral Cortex*. Springfield, IL: Charles Thomas.

Bostrom, N. (2014). *Superintelligence: Paths, Dangers, Strategies*. Oxford: Oxford University Press.

Boyden, E. S., F. Zhang, E. Bamberg, G. Nagel, and K. Deisseroth. (2005). "Millisecond-Timescale, Genetically Targeted Optical Control of Neural Activity." *Nature Neuroscience* 8: 1263–1268.

Brent, L. J. N., S. R. Heilbronner, J. E. Horvath, J. Gonzalez-Martinez, A. Ruiz-Lambides, A. G. Robinson, J. H. P. Skene, and M. L. Platt. (2013). "Genetic Origins of Social Networks in Rhesus Macaques." *Scientific Reports* 3.

Briggs, A. W., J. M. Good, R. E. Green, J. Krause, T. Maricic, U. Stenzel, C. Lalueza-Fox, P. Rudan, D. Brajković, Ž. Kućan, I. Gušić, R. Schmitz, V. B. Doronichev, L. V. Golovanova, M. de la Rasilla, J. Forea, A. Rosas, and S. Pääbo. (2009). "Targeted Retrieval and Analysis of Five Neanderthal mtDNA Genomes." *Science* 325 (5938): 318–321.

Broca, P. (1863). "Localization des functions cerebrales: Siege du langage articule." *Bulletins de la Societe d'Anthropologie (Paris)* 4: 200–203.

Brodal, A. (1981). *Neurological Anatomy in Relation to Clinical Medicine*. Oxford: Oxford University Press.

Brodmann, K. (1909). *Vergleichende Lokalisationslehre der Grosshirnrinde in ithren Prinzipien dargestellt auf Grund des Zellenbaues*. Leipzig: Barth.

Brown, J., J. I. Bianco, J. J. McGrath, and D. W. Eyles. (2003). "1,25 D3 Induces Nerve Growth Factor, Promotes Neurite Outgrowth and Inhibits Mitosis in Embryonic Rat Hippocampal Neurons." *Neuroscience Letters* 343: 139–143.

Brown, P., and C. D. Marsden. (1998). "What Do the Basal Ganglia Do?" *Lancet* 351: 1801–1804.

Buzsaki, G. (2006). *Rhythms of the Brain*. Oxford: Oxford University Press.

Byrne, R. W., and N. Corp. (2004). "Neocortex Size Predicts Deception Rate in Primates." *Proceedings of the Royal Society of London* 271: 1693–1699.

Cacioppo, J. T., P. S. Visser, and C. L. Pickett, eds. (2006). *Social Neuroscience: People Thinking About People*. Cambridge, MA: MIT Press.

Cajal, S. R. ([1906]) 1967. "The Structure and Connexions of Neurons." In *Nobel Lectures: Physiology or Medicine, 1901–1921,* 220–253. New York: Elsevier.

Cajal, S. R. ([1916] 1999). *Advice for a Young Investigator.* Cambridge, MA: MIT Press.

Calvert, G. A., E. T. Bullmore, M. J. Brammer, R. Campbell, S. C. R. Williams, P. K. McQuire, P. W. R. Woodruff, S. D. Iverson, and A. S. David. 1997. "Activation of Auditory Cortex During Silent Lipreading." *Science* 276 (5312): 593–596.

Cameron, O. G. (2002). *Visceral Sensory Neuroscience: Interoception.* Oxford: Oxford University Press.

Cannon, W. B. ([1915] 1963). *Bodily Changes in Pain, Hunger, Fear, and Rage.* New York: Harper and Row.

Cannon, W. B. ([1932] 1963). *The Wisdom of the Body.* New York: Norton.

Cardinal, R. N., J. A. Parkinson, J. Hall, and B. J. Everitt. (2002). "Emotion and Motivation: The Role of the Amygdala, Ventral Striatum and Prefrontal Cortex." *Neuroscience and Biobehavioral Reviews* 26 (3): 321–352.

Carter, C. S., L. Ahnert, K. E. Grossman, S. B. Hrdy, M. E. Lamb, S. W. Porgetes, and N. Sachser, eds. (2005). *Attachment and Bonding: A New Synthesis.* Cambridge, MA: MIT Press.

Carter, M., and L. de Lecea. (2011). "Optogenetic Investigation of Neural Circuits *in vivo.*" *Trends in Molecular Medicine* 17 (4): 197–206.

Carter, M., O. Yizhar, S. Chikahisa, H. Nguyen, A. Adamantidis, S. Nishino, K. Deisseroth, and L. de Lecea. (2010). "Tuning Arousal with Optogenetic Modulation of Locus Coeruleus Neurons." *Nature Neuroscience* 13 (12): 1526–1533.

Cassidy, A. S., P. Merolla, J. V. Arthur, S. K. Esser, B. Jackson, R. Alvarez-Icaza, P. Datta, J. Sawada, T. M. Wong, V. Feldman, A. Amir, D. B. Rubin, F. Akopyan, E. McQuinn, W. P. Risk, and D. S. Modha. (2013). "Cognitive Computing Building Block: A Versatile and Efficient Digital Neuron Model for Neurosynaptic Cores." 2013 International Joint Conference on Neural Networks, Dallas, TX. doi: 10.1109/IJCNN.2013 .6707077.

Catani, M., and S. Sandrone. (2015). *Brain Renaissance: From Vesalius to Modern Neuroscience.* Oxford: Oxford University Press.

Chalmers, D. (2003). "Consciousness and Its Place in Nature." In *The Black-well Guide to Philosophy of Mind*, ed. S. P Stich and T. A. Warfield, 102–142. Malden, MA: Blackwell.

Charney, D. S., E. J. Nestler, P. Sklar, and J. D. Buxbaum, eds. (2013). *Neurobiology of Mental Illness*. 4th ed. Oxford: Oxford University Press.

Chen, Y., and Z. A. Knight. (2016). "Making Sense of the Sensory Regulation of Hunger Reurons. *Bioessays* 38 (4): 1–8.

Cheney, D. L., and R. M. Seyfarth. (2007). *Baboon Metaphysics: The Evolution of a Social Mind*. Chicago: University of Chicago Press.

Chiao, J. Y., T. Iidaka, H. L. Gordon, J. Nogawa, M. Bar, E. Aminoff, N. Sadato, and N. Ambady. (2008). "Cultural Specificity in Amygdala Responses to Fear Faces." *Journal of Cognitive Neuroscience* 20 (12): 2167–2174.

Choleris, E., Little, S. R., Mong, J. A., Puram, S. V., Langer, R., and Pfaff, D. W. (2007). "Microparticle-Based Delivery of Oxytocin Receptor Antisense DNA in the Medial Amygdala Blocks Social Recognition in Female Mice." *Proceedings of the National Academy of Sciences of the United States*, 104: 4670–4675.

Chung, K., and K. Deisseroth. (2013). "CLARITY for Mapping the Nervous System." *Nature Methods* 10 (10): 513–520.

Chung, K., J. Wallace, S.-Y. Kim, S. Kalyanasundaram, A. S. Andalman, T. J. Davidson, J. J. Mirzabekov, K. A. Zalocusky, J. Mattis, A. K. Denisin, S. Pak, H. Bernstein, C. Ramakrishnan, L. Grosenick, V. Gradinaru, and K. Deisseroth. (2013). "Structural and Molecular Interrogation of Intact Biological Systems." *Nature* 497: 332–337.

Churchland, P. S., and T. J. Sejnowski. (2016). "Blending Computational and Experimental Neuroscience." *Nature Reviews Neuroscience* 17: 667–668.

Clark, A. (1998). *Being There: Bringing Brain, Body, and World Together*. Cambridge, MA: MIT Press.

Clark, A. (2003). *Natural-Born Cyborgs: Minds, Technologies, and the Future of Human Intelligence*. Oxford: Oxford University Press.

Clark, A. (2013). "Whatever Next? Predictive Brains, Situated Agents, and the Future of Cognitive Science." *Behavioral and Brain Sciences* 36: 1–73.

Cohen, M. S., S. M. Kosslyn, H. C. Breiter, G. J. DiGirolamo, W. L. Thompson, A. K. Anderson, S. Y. Brookheimer, B. R. Rosen, and J. W. Belliveau.

(1996). "Changes in Cortical Activity During Mental Rotation: A Mapping Study Using Functional MRI." *Brain* 119: 89–100.

Cohen, Z. D., and R. J. DeRubeis. (2018). "Treatment Selection in Depression." *Annual Review of Clinical Psychology* 14: 209–236.

Conlon, J. M., and D. Larhammar. (2005). "The Evolution of Neuroendocrine Peptides." *General and Comparative Endocrinology* 142: 53–59.

Connor, Steve. (1998). "Professor Has World's First Silicon Chip Implant." *The Independent*, August 26, 1998. https://www.independent.co.uk/news/professor-has-worlds-first-silicon-chip-implant-1174101.html.

Corlett, P. R., C. D. Frth, and P. C. Fletcher. (2009). "From Drugs to Deprivation: A Bayesian Framework for Understanding Models of Psychosis." *Psychopharmacology* 5: 515–530.

Cosmides, L., and J. Tooby. (1992). "Cognitive Adaptations for Social Exchange." In *The Adapted Mind: Evolutionary Psychology and the Generation of Culture*, ed. J. H. Barkow, J. Tooby, and L. Cosmides, 163–228. New York: Oxford University Press.

Coughlin, J., and R. Baran. (1995). *Neural Computation in Hopfield Networks and Boltzmann Machines*. Dover: University of Delaware Press.

Cowan, W., and E. Kandel. (2001). "A Brief History of Synapse and Synaptic Transmission." In *Synapses*, ed. W. M. Cowan, T. C. Sudhof, and C. F. Stevens, 1–88. Baltimore: Johns Hopkins University.

Crews, D. (2011). "Epigenetic Modifications of Brain and Behavior: Theory and Practice." *Hormones and Behavior* 59: 393–398.

Cristakis, N., and J. H. Fowler. (2007). "The Spread of Obesity in a Large Social Network Over 32 Years." *New England Journal of Medicine* 357: 370–379.

Curley, J. P., S. Davidson, P. Bateson, and F. A. Champagne. (2009). Social Enrichment During Postnatal Development Induces Transgenerational Effects on Emotional and Reproductive Behavior in Mice." *Frontiers in Behavioral Neuroscience* 3: 25.

Curley, J. P., and E. B. Keverne. (2005). "Genes, Brains and Mammalian Social Bonds." *Trends in Ecology and Evolution* 20: 561–567.

Damasio, A. R. (1994). *Descartes' Error: Emotion, Reason, and the Human Brain*. New York: Grosset/Putnam.

Damasio, A. R. (1996). "The Somatic Marker Hypothesis and the Possible Functions of the Prefrontal Cortex." *Philosophical Transactions of the Royal Society of London* 354: 1413–1420.

Darwin, C. ([1859] 1958). *The Origin of Species*. New York: Mentor.

Darwin, C. ([1871] 1874). *The Descent of Man and Selection in Relation to Sex*. Chicago: Rand, McNally.

Darwin, C. ([1872] 1965). *The Expression of Emotions in Man and Animals*. Chicago: University of Chicago Press.

Davidson, R. J., K. M. Putnam, and C. L. Larson. (2000). "Dysfunction in the Neural Circuitry of Emotion Regulations—A Possible Prelude to Violence." *Science* 289 (5479): 591–594.

Davis, M., and P. J. Whalen. (2001). "The Amygdala: Vigilance and Emotion." *Molecular Psychiatry* 6: 13–34.

Dawkins, R. (1996). *The Blind Watchmaker: Why the Evidence of Evolution Reveals a Universe Without Design*. New York: Norton.

Decety, J., and P. W. Jackson. (2006). "A Social Neuroscience Perspective on Empathy." *Current Directions in Psychological Science* 15: 54–58.

Decety, J., D. Perani, and M. Jeannerod. (1994). "Mapping Motor Representations with Positron Emission Tomography." *Nature* 371: 600–602.

Defelipe, J., and E. G. Jones. (1988). *Cajal on the Cerebral Cortex: An Annotated Translation of the Complete Writings*. Oxford: Oxford University Press.

Defense Advanced Research Projects Agency (DARPA). n.d. "Bridging the Bio-Electronic Divide." https://www.darpa.mil/about-us/bridging-the -bio-electronic-divide.

Defense Advanced Research Projects Agency (DARPA). n.d. "Targeted Neuroplasticity Training (TNT)." https://www.darpa.mil/program /targeted-neuroplasticity-training.

Dehaene, S., H. Lau, and S. Koudler. (2017). "What Is Consciousness and Could Machines Have It?" *Science* 358: 486–492.

Delgado, J. (1971). *Physical Control of the Mind: Toward a Psychocivilized Society*. New York: Harper and Row.

Dennett, D. (2017). *From Bacteria to Bach and Back: The Evolution of Minds*. New York: Norton.

Denton, D. A. (1982). *The Hunger for Salt: An Anthropological, Physiological and Medical Analysis*. New York: Springer-Verlag.

Denton, D. A. (2005). *The Primordial Emotions: The Dawning of Consciousness*. Oxford: Oxford University Press.

Denver, R. J. (2009). "Structural and Functional Evolution of Vertebrate Neuroendocrine Stress Systems." *Trends in Comparative Endocrinology and Neurobiology* 11: 1–16.

Descartes, R. ([1664] 2003). *Traite de l'homme*. Amherst, NY: Prometheus.

Desimone, R. (1996). "Neural Mechanisms for Visual Memory and Their Role in Attention." *Proceedings of the National Academy of Sciences* 93: 13494–13499.

Dethier, V. G. (1964). "Microscopic Brains." *Science* 143 (3611): 1138–1145.

Devlin, H. (2015). "Monkey 'Brain Net' Raises Prospects of Human Brain-to-Brain Communication." *Guardian*, July 9, 2015. https://www.the guardian.com/science/2015/jul/09/monkey-brain-net-raises-prospect-of -human-brain-to-brain-connection.

Dewey, J. (1896). "The Reflex Arc Concept in Psychology." *Psychological Review* 3 (4): 357–370.

Dewey, J. ([1910] 1965). *The Influence of Darwin on Philosophy*. Bloomington: Indiana University Press.

Dewey, J. ([1920] 1948). *Reconstruction in Philosophy*. Boston: Beacon.

de Waal, F., and F. Lanting. (1997). *Bonobo: The Forgotten Ape*. Berkeley: University of California Press.

Diamond, A. (2001). "A Model System for Studying the Role of Dopamine in the Prefrontal Cortex During Early Development of Humans: Early and Continuously Treated Phenylketonuria." In *Handbook of Developmental Cognitive Neuroscience*, ed. C. A. Nelson and M. Luciana, 433–472. Cambridge, MA: MIT Press.

Diamond, A., and D. Amaso. (2008). "Contributions of Neuroscience to Our Understanding of Cognitive Development." *Current Directions in Psychological Science* 17: 136–141.

Dierig, S. (2000). "Urbanization, Place of Experiment, and How the Electric Fish Was Caught by Emil du Bois-Reymond." *Journal of the History of the Neurosciences: Basic and Clinical Perspective* 9: 5–13.

Dobzhansky, T. C. (1962). *Mankind Evolving: The Evolution of the Human Species*. New Haven, CT: Yale University Press.

Dolan, R. (2007). "The Human Amygdala and Orbital Prefrontal Cortex in Behavioral Regulation." *Philosophical Transactions of the Royal Society of London, Series B: Biological Sciences* 362: 787–789.

Donald, M. (2001). *A Mind So Rare: The Evolution of Human Consciousness*. New York: Norton.

Donaldson, H. H. (1915). "The Rat: Data and Reference Tables for the Albino and Norway Rat." In *Memoirs of the Wistar Institute of Anatomy and Biology, no. 6*. Philadelphia: Wistar Institute.

Donaldson, Z. R., and L J. Young. (2008). "Oxytocin, Vasopressin, and the Neurogenetics of Sociality." *Science* 322: 900–904.

du Bois-Reymond, E. (1884). *Untersuchungen uber thierische elektricitat.* Berlin: Reimer.

Dum, R. P., and P. L. Strick. (2002). "Motor Areas in the Frontal Lobe of the Primate." *Physiology and Behavior* 77: 677–682.

Dunbar, R. I. M. (1992). "Neocortex Size as a Constraint on Group Size in Primates." *Journal of Human Evolution* 22: 469–493.

Dunbar, R. I. M. (2016). *Human Evolution: Our Brains and Behavior.* Oxford: Oxford University.

Dunbar, R. I. M., and S. Shultz. (2007). "Understanding Primate Evolution." *Philosophical Transactions of the Royal Society of London, Series B: Biological Sciences* 362: 649–658.

Eaglen, M. (2016). "What Is the Third Offset Strategy?" *Real Clear Defense,* February 15, 2016. https://www.realcleardefense.com/articles/2016/02/16/what_is_the_third_offset_strategy_109034.html.

Eccles, J. C. (1964). *The Physiology of Synapses.* New York: Academic Press.

Eichenbaum, H., and N. J. Cohen. (2001). *From Conditioning to Conscious Recollection: Memory Systems of the Brain.* Oxford: Oxford University Press.

Elliott, R., J. L. Newman, O. A. Longe, and J. F. W. Deakin. (2003). "Differential Response Patterns in the Striatum and Orbitofrontal Cortex to Financial Reward in Humans: A Parametric Functional Magnetic Resonance Imaging Study." *Journal of Neuroscience* 23: 303–307.

Ellman, J., L. Samp, and G. Coll. (2017). *Assessing the Third Offset Strategy.* Washington, DC: Center for Strategic and International Studies.

Emerging Technology from the arXiv. 2018. "The First 'Social Network' of Brains Lets Three People Transmit Thoughts to Each Other's Heads." *Technology Review,* September 29, 2018. https://www.technologyreview.com/s/612212/the-first-social-network-of-brains-lets-three-people-transmit-thoughts-to-each-others-heads/.

Emery, N., and D. Amaral. (2000). "The Role of the Amygdala in Primate Social Cognition." In *Cognitive Neuroscience of Emotion,* ed. R. Lane and L. Nadel. New York: Oxford University Press.

Enard, W., S. Gehre, K. Hammerschmidt, S. M. Hölter, T. Blass, M. Somel, . . . S. Pääbo. (2009). "A Humanized Version of Foxp2 Affects Cortico-Basal Ganglia Circuits in Mice." *Cell* 137: 961–971.

Engel, A. K., K. J. Friston, and D. Kragic. (2015). *The Pragmatic Turn: Towards Action-Oriented Views in Cognitive Science.* Cambridge, MA: MIT Press.

Evans, J. W., N. Lally, L. An, N. Li, A. C. Nugent, D. Banerjee, S. Snider, J. Shen, J. P. Roiser, and C. A. Zarate Jr.. (2018). "7T ¹H-MRS in Major Depression Disorder: A Ketamine Treatment Study." *Neuropharmacology* 43: 1908–1914.

Falk, D. (1983). "Cerebral Cortices of East African Early Hominids." *Science* 221: 1072–1074.

Falk, E., and C. Scholz. (2018). "Persuasion, Influence, and Value: Perspectives From Communication and Social Neuroscience." *Annual Review of Psychology* 69: 329–356.

Farah, M. J. (1984). "The Neurobiological Basis of Visual Imagery: A Componential Analysis." *Cognition* 18: 245–272.

Farah, M. J. (2018). "Socioeconomic Status and the Brain: Prospects for Neuroscience Informed Policy." *Nature Reviews Neuroscience* 19: 428–438.

Ferguson, N. (2018). *The Square and the Tower: From the Freemasons to Facebook.* New York: Random House.

Finger, S. (1994). *Origins of Neuroscience: A History of Explorations into Brain Function.* Oxford: Oxford University Press.

Finger, S. (2000). *Minds Behind the Brain: A History of the Pioneers and Their Discoveries.* Oxford: Oxford University Press.

Fitzsimons, J. T. (1999). "Angiotensin, Thirst, and Sodium Appetite." *Physiological Reviews* 76: 383–366.

Fiorillo, C. D., P. N. Tobler, and W. Schultz. (2003). "Discrete Coding of Reward Probability and Uncertainty by Dopamine Neurons." *Science* 299: 1898–1902.

Fluharty, S. J., and A. N. Epstein. (1983). "Sodium Appetite Elicited by Intracerebroventricular Infusion of Angiotensin II in the Rat." *Behavioral Neuroscience* 97: 746–758.

Flynn, J. P. (1972). "Patterning Mechanisms, Patterned Reflexes, and Attack Behavior in Cats." *Nebraska Symposium on Motivation* 20: 125–153.

Foley, R. (2001). "In the Shadow of the Modern Synthesis?" *Evolutionary Anthropology* 10: 5–14.

Fregnac, Y. (2017). "Big Data and the Industrialization of Neuroscience: A Safe Roadmap for Understanding the Brain?" *Science* 358: 470–477.

Freud, S. ([1900] 1976). *The Interpretation of Dreams*. New York: Collier.

Friston, K. J., M. Lin, C. D. Frith, and G. Pezzulo. (2017). "Active Inference, Curiosity, and Insight." *Neural Computation* 29: 1–49

Frith, C. D. (2007). "The Social Brain?" *Philosophical Transactions of the Royal Society of London* 362: 671–678.

Frith, C. D., and D. Wolpert, ed. (2004). *The Neuroscience of Social interaction: Decoding, Imitating, and Influencing the Actions of Others*. Oxford: Oxford University Press.

Fullam, R. S., S. McKie, and M. C. Dolan. (2009). "Psychopathic Traits and Deception: Functional Magnetic Resonance Imaging Study." *British Journal of Psychiatry* 194: 229–235.

Fulton, J. F. (1949). *Functional Localization in the Frontal Lobes and Cerebellum*. Oxford: Oxford University Press.

Fuster, J. M. (2003). *Cortex and Mind: Unifying Cognition*. Oxford: Oxford University Press.

Galileo ([1610] 1957). "The Starry Messenger." In *Discoveries and opinions of Galileo*, ed. S. Drake. New York: Doubleday.

Gall, E. J. ([1822–1899] 1935). *On the Function of the Brain and Each of Its Parts*. Boston: Marsh, Capen and Lyon.

Gallagher, M., and F. C. Holland. (1994). "The Amygdala Complex: Multiple Roles in Associative Learning and Emotion." *Proceedings of the National Academy of Sciences of the United States of America* 91: 11771–11776.

Gallistel, C. R. (1980). *The Organization of Action: A New Synthesis*. Hillsdale, NJ: Lawrence Erlbaum.

Gallistel, C. R. (1990). *The Organization of Learning*. Cambridge, MA: MIT Press.

Garrett, N., S. C. Lazzaro, and T. Sharot. (2016). "The Brain Adapts to Dishonesty." *Nature Neuroscience* 19: 1717–1732.

Gazzaniga, M. S. (1985). *The Social Brain*. New York: Basic Books.

Gazzaniga, M. S. ([1995] 2015). *The New Cognitive Neurosciences*. Cambridge, MA: MIT Press.

Geschwind, N. (1974). *Selected Papers on Language and the Brain*. Boston: Reidel.

Gibson, D. G., J. I. Glass, C. Lartigue, V. N. Noskov, . . . J. Craig Venter. 2010. "Creation of a Bacterial Cell Controlled by a Chmically Synthesized Genome." *Science* 329 (5987): 52–56. DOI:10.1126/science.1190719.

Gibson, J. J. (1966). *The Senses Considered as Perceptual Systems.* New York: Houghton-Mifflin.

Gigerenzer, G. (2000). *Adaptive Thinking: Rationality in the Real World.* New York: Oxford University Press.

Gimpl, G., and F. Fahrenholz. (2001). "The Oxytocin Receptor System: Structure, Function, and Regulation." *Physiological Reviews* 81: 629–683.

Glimcher, P. W. (2003). *Decisions, Uncertainty, and the Brain: The Science of Neuroeconomics.* Cambridge, MA: MIT Press.

Goldman-Rakic, P. S. (1996). "The Prefrontal Landscape: Implications of Functional Architecture for Understanding Human Mentation and the Central Executive." *Philosophical Transactions Royal Society of London* 351: 1445–1453.

Goldman-Rakic, P. S., C. Leranth, S. M. Williams, N. Mons, and M. Gerrard. (1989). "Dopamine Synaptic Complex with Pyramidal Neurons in Primate Cerebral Cortex." *Proceedings of the National Academy of Sciences of the United States of America* 86: 9015–9019.

Goldstein, K. (1939). *The Organism.* New York: American Book Company.

Golgi, Camillo. (1906). "The Neuron Doctrine: Theory and Facts." Nobel lecture, December 11, 1906.

Goodman, N. ([1955] 1978). *Fact, Fiction, and Forecast.* New York: Bobbs-Merill.

Goodson, J. L. (2005). "The Vertebrate Social Behavioral Network: Evolutionary Themes and Variations." *Hormones and Behavior* 48: 11–22.

Gordon, A. D., L. Nevell, and B. Wood. (2008). "The *Homo floresiensis* Cranium (LB1): Size, Scaling, and Early *Homo* Affinities." *Proceedings of the National Academy of Sciences of the United States of America* 105: 4650–4655.

Gotts, S., K. Simmons, L. Milbury, G. Wallace, R. Cox, and A. Martin. (2012). "Fractionation of Social Brain Circuits in Autism Spectrum Disorders." *Brain* 135: 2711–2725.

Gould, E., B. S. McEwen, P. Tanapat, L. A. Galea, and E. Fuchs. (1996). "Neurogenesis in the Dentate Gyrus of the Adult Tree Shrew Is Regulated by Psychosocial Stress and NMDA Receptor Activation." *Journal of Neuroscience* 17: 2492–2498.

Gould, E., N. Vall, M. Wagers, and C. G. Gross. (2001). "Adult Generated Hippocampal and Neocortical Neurons in Macaques Have a Transient

Existence." *Proceedings of the National Academy of Sciences of the United States of America* 98: 101910–101916.

Gould, S. J. (2002). *The Structure of Evolutionary Theory.* Cambridge, MA: Harvard University Press.

Graybiel, A. (1998). "The Basal Ganglia and Chunking of Action Repertoires." *Neurobiology of Learning and Memory* 70: 119–136.

Greely, H. T. (2011). "Reading Minds with Neuroscience: Possibilities for the Law." *Cortex* 47: 1254–1255.

Greely, H. T., and J. Illes. (2007). "Neuroscience-Based Lie Detection: The Urgent Need for Regulation." *American Journal of Law and Medicine* 33: 377–431.

Greenbough, W., and F. Volkmar. (1974). "Pattern of Dendritic Branching in Occipital Cortex of Rats Reared in Complex Environments." *Experimental Neurobiology* 40: 491–504.

Greene, J. D., and J. M. Paxton. (2009). "Patterns of Neural Activity Associated with Honest and Dishonest Moral Decision." *Proceedings of the National Academy of Sciences of the United States of America* 106: 12506–12511.

Greene, J. D., R. B. Sommerville, L. E. Nystrom, J. M. Darley, and J. D. Cohen. (2001). "An fMRI Investigation of Emotional Engagement in Moral Judgment." *Science* 293: 2105–2108.

Greene, J. D., S. A. Morelli, K. Lowenberg, L. E. Nystrom, and J. D. Cohen. (2008). "Cognitive Load Selectively Interferes with Utilitarian Moral Judgment." *Cognition* 107 (3): 1144-54.

Greene, J. D. (2013). *Moral Tribes: Emotion, Reason, and the Gap Between Us and Them.* New York: Penguin.

Greene, J. D. (2014). "Beyond Point-and-Shoot Morality: Why Cognitive Neuroscience Matters for Ethics." *Ethics* 124: 695–726.

Gregory, R. L. (1973). *Eye and Brain: The Psychology of Seeing.* New York: McGraw-Hill.

Griffin, D. M., D. S. Hoffman, and P. L. Strick. (2015). "Corticomotorneuronal Cells Are Functionally Tuned." *Science* 350: 667–670.

Griffin, D. R. (1958). *Listening in the Dark: The Acoustic Orientation of Bats and Men.* New Haven, CT: Yale University Press.

Grill, H. J. (2006). "Distributed Neural Control of Energy Balance: Contributions from Hindbrain and Hypothalamus." *Obesity* 14: 216S-221S.

Gross, C. G. (1998). *Brain, Vision, Memory: Tales in the History of Neuroscience.* Cambridge, MA: MIT Press.

Gründemann, J., Y. Bitterman, T. Lu, S. Krabbe, B. F. Grewe, M. J. Schnitzer, and A. Lüthi. (2019). "Amygdala Ensembles Encode Behavioral States." *Science* 364.

Hacking, I. (1999). *The Taming of Chance.* Cambridge: Cambridge University Press.

Haidt, J. (2007). "The New Synthesis in Moral Psychology." *Science* 316: 998–1002.

Hanson, N. R. ([1958] 1972). *Patterns of Discovery: An Inquiry into the Conceptual Foundations of Science.* Cambridge: Cambridge University Press.

Hare, B., V. Wobber, and R. Wrangham. (2012). "The Self-Domestication Hypothesis: Evolution of Bonobo Psychology Is Due to Selection Against Aggression." *Animal Behavior* 83: 573–585.

Harris, J. M., F. J. Brown, M. G. Leakey, A. C. Walker, and R. E. Leakey (1988). "Pliocene and Pleistocene Hominid-Bearing Sites from West of Lake Turkana, Kenya." *Science* 239: 27–32.

Hawrylycz, M. J., E. S. Lein, A. L. Guillozet-Bongaarts, E. H. Shen, L. Ng, J. A. Miller, . . . A. R. Jones. (2012). "An Anatomically Comprehensive Atlas of the Adult Brain Transcriptome." *Nature* 489: 391–399.

Hawrylycz, M., J. A. Miller, V. Menon, D. Feng, T. Dolbeare, A. L. Guillozet-Bongaarts, . . . E. Lein. (2016). "Canonical Genetic Signatures of the Adult Human Brain." *Nature Neuroscience* 18: 1832–1842.

Haxby, J. V., M. I. Gobbini, M. L. Furey, A. Ishai, J. L. Schouten, and P. Pietrini. (2001). "Distributed and Overlapping Representations of Faces and Objects in Ventral Temporal Cortex." *Science* 293: 2425–2430.

Haxby, J. V., E. A. Hoffman, and M. I. Gobbini. (2000). "The Distributed Human Neural System for Face Perception." *Trends in Cognitive Sciences* 4: 223–233.

Hayden, B. Y., J. M. Pearson, and M. L. Platt. 2011. "Neuronal Basis of Sequential Foraging Decisions in a Patchy Environment." *Nature Neuroscience* 14 (7): 933-939.

Hebb, D. O. (1949). *The Organization of Behavior: A Neuropsychological Theory.* New York: Wiley.

Heinrichs, M., and G. Domes. (2008). "Neuropeptides and Social Behaviour: Effects of Oxytocin and Vasopressin in Humans." *Progress in Brain Research* 170: 337–350.

Hill, M. J., R. J. McLaughlin, B. Bingham, L. Shrestha, T. T. Y. Lee, J. M. Gray, C. J. Hillard, B. B. Gorzalka, and V. Viau. (2010). "Endogenous Cannabinoid Signaling Is Essential for Stress Adaptation." *Proceedings of the National Academy of Science* 107: 9406–9411.

Helmholtz, H. von. ([1867] 1963). *Handbook of Physiological Optics.* New York: Dover.

Herbert, J. (1993). "Peptides in the Limbic System: Neurochemical Codes for Co-Ordinated Adaptive Responses to Behavioral and Physiological Demand." *Progress of Neurobiology* 41: 723–791.

Herbert, J. (2015). *Testosterone: Sex, Power, and the Will to Win.* Oxford: Oxford University Press.

Herbert, J., and P. J. Lucassen. (2016). "Depression as a Risk Factor for Alzheimer's Disease: Genes, Steroids, Cytokines, and Neurogenesis— What Do We Need to Know?" *Frontiers in Neuroendocrinology* 41: 153–171.

Herculano-Houzel, S. (2016). *The Human Advantage: A New Understanding for How Our Brains Became Remarkable.* Cambridge, MA: MIT Press.

Herrick, C. J. (1905). "The Central Gustatory Pathway in the Brain of Body Fishes." *Journal of Comparative Neurology* 15: 375–486.

Herrick, C. J. ([1926] 1963). *Brain in Rats and Men: A Survey of the Origin and Biological Significance of the Cerebral Cortex.* New York: Hafner.

Hess, W. R. ([1948] 1957). *The Functional Organization of the Diencephalon.* New York: Orange and Stratton.

Hirstein, W., J. Poland, and J. Radden. (2005). *Brain Fiction: Deception and the Riddle of Confabulation.* Cambridge, MA: MIT Press.

Hirstein, W., K. L. Stifford, and T. K. Fagan, (2018). *Responsible Brains: Neuroscience, Law, and Human Culpability.* Cambridge, MA: MIT Press.

Hoag, H. (2003). "Neuroengineering Remote Control." *Nature* 423: 796–798.

Hobson, J. A. (1988). *The Dreaming Brain.* New York: Basic Books.

Hodos, W., and A. B. Butler. (1997). "Evolution of Sensory Pathways in Vertebrates." *Brain, Behavior and Evolution* 50: 189–199.

Hofer, M. A., and R. M. Sullivan. (2001). "Toward a Neurobiology of Attachment." In *Handbook of Developmental Cognitive Neuroscience,* ed. C. N. Nelson and M. Luciana, 599–616. Cambridge, MA: MIT Press.

Hokfelt, T., J. Fahrenkrug, K. Tatemoto, V. Mutt, S. Werner, A. L. Hulting, . . . K. J. Chang. (1983). "The PHI (PHI-27)/Corticotropin-Releasing

Factor/Enkephalin Immunoreactive Hypothalamic Neuron: Possible Morphological Basis for Integrated Control of Prolactin, Corticotropin, and Growth Hormone Secretion." *Proceedings of the National Academy of Sciences of the United States of America* 80: 895–898.

Holland, L. Z., and S. Short. (2008). "Gene Duplication, Co-Option, and Recruitment During the Origin of the Vertebrate Brain From the Invertebrate Chordate Brain." *Brain Behavior and Evolution* 72: 91–105.

Hollander, E., J. Bartz, W. Chaplin, A. Phillips, J. Sumner, L. Soorya, . . . S. Wasserman. (2006). "Oxytocin Increases Retention of Social Cognition in Autism." *Biological Psychiatry* 61: 493–503.

Holliday, R. (2002). "Epigenetics Comes of Age in the Twenty-First Century." *Journal of Genetics* 81: 1–4.

Holloway, M. (1985). "Size and Shape of the Cerebral Cortex in Mammals: The Cortical Surface." *Brain, Behavior, and Evolution* 27: 28–40.

Holst, E. von. (1973). "Relative Coordination as a Phenomenon and as a Method of Analysis of Central Nervous Functions." In *The Behavioral Physiology of Animals and Man: Selected Papers of Erich von Holst.* Coral Gables, FL: University of Miami Press.

Holst, E. von, and U. von St. Paul. (1963). "On the Functional Organization of Drives." *Animal Behaviour* 11: 1–20.

Holtzheimer, P., and H. Mayberg. (2011). "Deep Brain Stimulation for Psychiatric Disorders." *Annual Review of Neuroscience* 34: 289–307.

Horikawa, T., M. Tamaki, Y. Miyawaki, and Y. Kamitani. 2013. "Neural Decoding of Visual Imagery During Sleep." *Science* 340 (6132): 639–642.

Horn, S. R., D. S. Charney, and A. Feder. (2016). "Understanding Resilience: New Approaches for Preventing and Treating PTSD." *Experimental Neurology* 284: 119–132.

Hubel, D. H., and T. N. Wiesel. (2005). *Brain and Visual Perception: The Story of a 25-Year Collaboration.* Oxford: Oxford University Press.

Hubel, D. H., T. N. Wiesel, and S. LeVay. (1977). "Plasticity of Ocular Dominance Columns to Monkey Striate Cortex." *Philosophical Transactions of the Royal Society of London, Series B: Biological Sciences* 278: 377–409.

Humphrey, N. (1976). "The Social Function of Intellect." In *Growing Points in Ethology*, ed. P. P. G. Bateson and R. A. Hinde, 307–317. Cambridge: Cambridge University Press.

Huth, A.G., T. Lee, S. Nishimoto, N. Y. Bilenko, A. T. Vu, and J. L. Gallant. (2016). "Decoding the Semantic Content of Natural Movies from Human Brain Activity." *Frontiers in Systems Neuroscience* 10: 81-91.

Hwang, T., C.-K. Park, A. K. L. Leung, Y. Gao, T. M. Hyde, J. E. Kleinman, A. Rajpurohit, R. Tao, J. H. Shin, and D. R. Weinberger. (2016). "Dynamic Regulation of RNA Editing in Human Brain Development and disease. *Nature Neuroscience* 19: 1093-1099.

Iacoboni, M., J. Freedman, and J. Kaplan. (2007). "This Is Your Brain on Politics." *New York Times*, November 11, 2007.

Iacoboni, M., R. P. Woods, M. Brass, H. Bekkering, J. C. Mazziotta, and G. Rizzolatti. (1999). "Cortical Mechanisms of Imitation." *Science* 286: 2526-2528.

Ikemoto, S., and J. Panksepp. (1999). "The Role of Nucleus Accumbens Dopamine in Motivated Behavior: A Unifying Interpretation with Special Reference to Reward-Seeking." *Brain Research Reviews* 31: 6-41.

Illes, J. (2006). *Neuroethics: Anticipating the Future*. Oxford: Oxford University Press.

Insel, T. R., and R. D. Fernald. (2004). "How the Brain Processes Social Information." *Annual Review of Neuroscience* 27: 697-722.

Israel, S., E. Lerer, I. Shalev, F. Uzefovsky, M. Reibold, R. Bachner-Melman, . . . R. F. Ebstein. (2008). "Molecular Genetic Studies of the Arginine Vasopressin 1a Receptor (AVPR1a) and the Oxytocin Receptor (OXTR) in Human Behavior: From Autism to Altruism with Some Notes in Between." *Progress in Brain Research* 170: 435-449.

Jackson, J. H. ([1884] 1958). "Evolution and Dissolution of the Nervous System." In *Selected Writings of John Hughlings Jackson*. London: Staples.

Jackson, P. L., and J. Decety. (2004). "Motor Cognition: A New Paradigm to Self and Other Interactions." *Current Opinion in Neurobiology* 14: 259-263.

Jacobowitz, D. M. (2006). "The Birth of Neurochemical Maps." *Neurochemical Research* 31: 125-126.

James, W. ([1890] 1952). *The Principles of Psychology*. Vols. 1-2. New York: Henry Holt.

James, W. ([1896] 1956). *The Will to Believe, Human Immortality*. New York: Dover.

James, W. (1899). *Talks to Teachers on Psychology: And to Students on Some of Life's Ideals*. New York: Henry Holt.

James, W. ([1910] 1961). *The Varieties of Religious Experience*. New York: Collier.

Jaspers, K. ([1913] 1997). *General Psychopathology*. Vols. 1–2. Trans. J. Hoenig and M. W. Hamilton. Baltimore: Johns Hopkins University Press.

Jeannerod, M. (1997). *The Cognitive Neuroscience of Action*. Oxford: Blackwell.

Johnson, M. (2014). *Morality for Humans: Ethical Understanding from the Perspective of Cognitive Science*. Chicago: University of Chicago Press.

Johnson-Frey, S. H. (2003). "What's So Special About Tool Use?" *Neuron* 39: 201–204.

Jolly, A. (1999). *Lucy's Legacy: Sex and Intelligence in Human Evolution*. Cambridge, MA: Harvard University Press.

Kagan, J. (2002). *Surprise, Uncertainty, and Mental Structures*. Cambridge, MA: Harvard University Press.

Kahana, M. J. (2006). "The Cognitive Correlates of Human Brain Oscillations." *Journal of Neuroscience* 26: 1669–1672.

Kahneman, D., P. Slovic, and A. Tversky. (1982). *Judgment Under Uncertainty: Heuristics and Biases*. Cambridge: Cambridge University Press.

Kamienski, L. ([2012] 2016). *Shooting Up: A Short History of Drugs and War*. Oxford: Oxford University Press.

Kandel, E. R., J. Jessell, and J. H. Schwartz. ([1981] 2012). *Principles of Neural Science*. New York: McGraw-Hill.

Kandel, E. R., and L. R. Squire. (2000). "Neuroscience: Breaking Down Scientific Barriers to the Study of Brain and Mind." *Science* 290: 1113–1120.

Kanwisher, N. (2006). "What's in a Face?" *Science* 311: 617–618.

Kanwisher, N., J. McDermott, and M. Chun. (1997). "The Fusiform Face Area: A Module in Human Extrastriate Cortex Specialized for Face Perception." *Journal of Neuroscience* 17: 4302–4311.

Kaplan, H. S., and A. J. Robson. (2002). "The Emergence of Humans: The Coevolution of Intelligence and Longevity with Intergenerational Transfers." *Proceedings of the National Academy of Sciences of the United States of America* 99: 10221–10226.

Kappers, C. U. A., G. C. Huber, and E. C. Crosby. (1967). *The Comparative Anatomy of the Nervous System of Vertebrates, Including Man*. New York: Hafner.

Kempermann, G. (2006). *Adult Neurogenesis*. Oxford: Oxford University Press.

Kendrick, K. M. (2000). "Oxytocin, Motherhood, and Bonding." *Experimental Physiology* 85: 111S–124S.

Kerns, J. G., J. D. Cohen, A. W. MacDonald, R. Y. Cho, V. A. Stenger, and C. S. Carter. (2004). "Anterior Cingulate Conflict Monitoring and Adjustments in Control." *Science* 202: 1023–1026.

Keverne, E. B., and J. P. Curley. (2004). "Vasopressin, Oxytocin, and Social Behavior." *Current Opinion in Neurobiology* 14: 777–783.

Keverne, E. B., and J. P. Curley. (2008). "Epigenetics, Brain Evolution, and Behaviour." *Frontiers in Neuroendocrinology* 29: 398–412.

Kim, D. H. (2010). "Dissolvable Films of Silk Fibroin for Ultrathin, Conformal Bio-Integrated Electronics." *Nature Matter* 9: 511–517.

King, A. N., A. F. Barber, A. E. Smith, A. P. Dreyer, D. Sitaraman, M. N. Nitabach, D. J. Cavanaugh, and A. Sehgal. (2018). "A Peptidergic Circuit Links Circadian Clock to Locomotor Activity." *Current Biology* 27: 1915–1927.

Kirkby, L. A., F. J. Luongo, M. B. Lee, M. Nahum, T. M. Van Vleet, V. R. Rao, H. E. Dawes, E. F. Chang, and V. S. Sohal. (2018). "An Amygdala-Hippocampus Subnetwork That Encodes Variation in Human Mood." *Cell* 18: 31313–31318.

Kirsch, P., C. Esslinger, Q. Chen, D. Mier, S. Lis, S. Siddhanti, H. Gruppe, V. S. Mattay, B. Gallhofer, and A. Meyer-Lindenberg. (2005). "Oxytocin Modulates Neural Circuitry for Social Cognition and Fear in Humans." *Journal of Neuroscience* 25: 11489–11493.

Kissenger, Henry A. 2018. "How the Enlightenment Ends." *The Atlantic*, June 2018. https://www.theatlantic.com/magazine/archive/2018/06/henry-kissinger-ai-could-mean-the-end-of-human-history/559124/.

Kitcher, P. (1993). *The Advancement of Science: Science Without Legend, Objectivity Without Illusions*. Oxford: Oxford University Press.

Klein, R. G. (2008). "Out of Africa and the Evolution of Human Behavior." *Evolutionary Anthropology* 17: 267–281.

Kluver, H. M., and P. C. Bucy. (1939). "Preliminary Analysis of Functions of the Temporal Lobes in Monkeys." *Archives of Neurology and Psychiatry* 42: 979–1000.

Koch, C., and R. Reid. (2012). "Neuroscience: Observation of the Mind." *Nature* 483: 397–398.

Koch, S., M. van Zuiden, L. Nawijn, J. Frijling, D. Veltman, and M. Olff. (2014). "Intranasal Oxytocin as Strategy for Medication-Enhanced Psychotherapy of PTSD: Salience Processing and Fear Inhibition Processes." *Psychoneuroendocrinology* 40: 242–256.

Koechlin, E., and A. Hyafil. (2007)." Anterior Prefrontal Function and the Limits of Human Decision-Making." *Science* 318: 594–598.

Kolber, B. J., M. S. Roberts, M. P. Howell, D. F. Wozniak, M. S. Sands, and L. J. Muglia. (2008). "Central Amygdala Glucocorticoid Receptor Action Promotes Fear-Associated CRH Activation and Conditioning." *Proceedings of the National Academy of Sciences of the United States of America* 105: 12004–12009.

Konorski, J. (1967). *Integrative Activity of the Brain: An Interdisciplinary Approach.* Chicago: University of Chicago Press.

Koob, G. F. (2015). "The Dark Side of Emotion: The Addiction Perspective." *European Journal of Pharmacology* 15: 73–87.

Koob, G. F., and M. LeMoal. (2005). *Neurobiology of Addiction.* New York: Elsevier.

Kosfeld, M., M. Heinrichs, P. J. Zak, U. Fischbacher, and E. Fehr. (2005). "Oxytocin Increases Trust in Humans." *Science* 435, 673–676.

Kosslyn, S. (1984). *Ghosts in the Mind's Machine: Creating and Using Images in the Brain.* New York: Norton.

Kringelbach, M. T., and K. C. Berridge. (2010). *Pleasures of the Brain.* Oxford: Oxford University Press.

Kropotkin, P. ([1902] 1976) *Mutual Aid: A Factor in Evolution.* New York: Extending Horizions Book.

Kuffler, S., J. Nicholls, and A. Martin. (1984). *From Neuron to Brain.* Sunderland, MA: Sinauer.

Kuhn, T. S. (1962). *The Structure of Scientific Revolutions.* Chicago: University of Chicago Press.

Kuhn, T. S. (2000). *The Road Since Structure.* Chicago: University of Chicago Press.

Laberge, F., S. Mühlenbrock-Lenter, W. Grunwald, and G. Roth. (2006). "Evolution of the Amygdala: New Insights from Studies in Amphibians." *Brain, Behavior, and Evolution* 67: 177–187.

Landgraf, R., and I. Neuman. (2008). "Advances in Vasopressin and Oxytocin: From Genes to Behavior to Disease." *Progress in Brain Research, 170,* 435–449.

Langleben, D. D., J. W. Loughead, W. B. Bilker, K. Ruparel, A. R. Childress, S. I. Busch, and R. C. Gur. (2005). "Telling the Truth From Lie in Individual Subjects with Fast Event-Related fMRI." *Human Brain Mapping* 26: 262–272.

Langleben, D. D., and J. C. Moriarty. 2013. "Using Brain Imaging for Lie Detection: Where Science, Law and Research Policy Collide." *Psychology, Public Policy, and Law* 19 (2): 222–234.

Lashley, K. S. (1938). "An Experimental Analysis of Instinctive Behavior." *Psychological Review* 45: 445–471.

Lashley, K. S. (1960). *The Neuropsychology of Karl Lashley.* New York: McGraw-Hill.

Lazaridou, A., J. Kim, C. M. Cahalan, M. L. Loggia, O. Franceschelli, C. S. Berna, P. H. Schur, V. Napadow, and R. F. R. Edwards. (2017). "Effects of Cognitive-Behavioral Therapy (CBT) on Brain Connectivity Supporting Catastrophizing in Fibromyalgia." *Clinical Journal of Pain* 33: 215–221.

Lebrfrv, M. E., and M. A. L. Huang (2006). "Brain Machine Interfaces." *Trends in Neuroscience* 29: 536–546.

Leakey, L. S. B. ([1934] 1954). *Adam's Ancestors: The Evolution of Man and His Culture.* New York: Harper Torchbooks.

Leakey, M. G., F. Spoor, F. H. Brown, P. N. Gathogo, C. Kiarie, L. N. Leakey, and I. McDougall. (2001). "New Hominin Henus from Eastern Africa Shows Diverse Middle Pliocene Lineages." *Nature* 410: 433–440.

LeDoux, J. E. (2012). "Rethinking the Emotional Brain." *Neuron* 73: 653–676.

LeDoux, J. E. (2015). *Anxious: Using the Brain to Understand and Treat Fear and Anxiety.* New York: Viking.

Lee, H. M., P. M. Gigurere, and B. L. Roth. (2014). "DREADDs: Novel Tools for Drug Discovery and Development." *Drug Discovery Today* 19: 469–473.

Leigh, S. R. (2004). "Brain Growth, Life History, and Cognition in Primate and Human Evolution." *American Journal of Human Evolution* 62: 139–164.

Lein, E. S., M. J. Hawrylycz, N. Ao, M. Ayres, A. Bensinger, A. Bernard, . . .
A. R. Jones. (2007). "Genome-wide Atlas of Gene Expression in the
Adult Mouse Brain." *Nature* 445: 168–176.

Li, M., A. E. Jaffe, R. E. Straub, R. Tao, J. H. Shin, Y. Wang, . . . D. R.
Weinberger. (2016). "A Human-Specific AS3MT Isoform and BORC57
Are Molecular Risk Factors in the 10q24.32 Schizophrenia Associated
Locus." *Nature Medicine* 22: 649–656.

Lieberman, D. E. (2011). *The Evolution of the Human Head*. Cambridge, MA:
Harvard University Press.

Lieberman, P. (2000). *Human Language and Our Reptilian Brain: The Subcortical Bases of Speech, Syntax, and Thought*. Cambridge, MA: Harvard
University Press.

Lilly, J. C. (1967). *The Mind of the Dolphin: A Nonhuman Intelligence*. New
York: Doubleday.

Lindberg, J., S. Björnerfeldt, M. Bakken, C. Vilà, E. Jazin, and P. Saetre.
(2007). "Selection for Tameness Modulates the Expression of Heme
Related Genes in Silver Foxes." *Behavioral Brain Research* 3: 18–28.

Liu, Y., and Z. X. Wange. (2003). "Nucleus Accumbens Oxytocin and
Dopamine Interact to Regulate Pair Bond Formation in Female Prairie
Voles." *Neuroscience* 121: 537–544.

Lovejoy, D. A., and R. J. Balment. (1999). "Evolution and Physiology of the
Corticotropin-Releasing Factor (CRF) Family and Neuropeptides in
Vertebrates." *General & Comparative Endocinology* 115: 1–22.

Lowell, B. B. (2019). "New Neuroscience of Homeostasis and Drives
for Food, Water, and Salt." *New England Journal of Medicine* 3803:
459–571.

Lundy, R. F., and R. Norgren. (2004). "Gustatory System." In *The Rat Nervous System*, 3rd ed., ed. G. Paxnios, 890–920. San Diego, CA: Academic Press.

Luria, A. R. (1973). *The Working Brain: An Introduction to Neuropsychology*.
New York: Basic Books.

Maclean, P. D. (1990). *The Triune Brain in Evolution: Role in Paleocerebral
Functions*. New York: Plenium.

Magoun, H. (1958). *Waking Brain*. Springfield, IL: CC Thomas.

Markram, H., E. Muller, S. Ramaswamy, M. W. Reimann, M. Abdellah,
C. Aguado Sanchez, . . . F. Schürmann. (2015). "Reconstruction and
Simulation of Neocortical Microcircuitry." *Cell* 163: 456–492.

Marino, L., R. C. Connor, R. E. Fordyce, L. M. Herman, P. R. Hof, L. Lefebvre, . . . H. Whitehead. (2007). "Cetaceans Have Complex Brains for Complex Cognition." *PLoS Biology* 5: 966–972.

Marsden, C. D. (1984). "The Pathophysiology of Movement Disorders." *Neurologic Clinics* 2: 435–459.

Marsden, C. D., H. B. Merton, J. E. R. Adam, and M. Hallett. (1978). "Automatic and Voluntary Responses to Muscle Stretch in Man." In *Progress in Clinical Neurophysiology*, vol. 4, *Cerebral Motor Control in Man: Long Loop Mechanisms*, ed. J. E. Desmedt, 167–177. Basel: Karger.

Martin, A. (2007). "The Representation of Object Concepts in the Brain." *Annual Review of Psychology* 58: 25–45.

Martin, A., and J. Weisber. (2003). "Neural Foundations for Understanding Social and Mechanical Concepts." *Cognitive Neuropsychology* 20: 575–587.

May, M. T., trans. (1968). *Galen on the Usefulness of the Parts of the Body*. Vol. 1. Ithaca, NY: Cornell University Press.

Mayberg, H. (1997). "Limbic-Cortical Dysregulation: A Proposed Model of Depression." *Journal of Neuropsychiatry and Clinical Neurosciences* 9: 471–481.

Mayr, E. (1991). *One Long Argument: Charles Darwin and the Genesis of Modern Evolutionary Thought*. Cambridge, MA: Harvard University Press.

McCarthy, M. M. (2008). "Estradiol and the Developing Brain." *Physiological Reviews* 88: 91–134.

McClintock, B. (1951). "Chromosome Organization and Genic Expression." *Cold Spring Harbor Symposia on Quantitative Biology* 16: 13–47.

McCulloch, W. S. (1965). *Embodiments of Mind*. Cambridge, MA: MIT Press.

McDermott, D. (2007). "Artificial Intelligence and Consciousness." In *The Cambridge Handbook of Consciousness*, ed. P. D. Zelazo, M. Moscovitch, and E. Thompson, 117–150. New York: Cambridge University Press.

McEwen, B. S. (2007). "Physiology and Neurobiology of Stress and Adaptation: Central Role of the Brain." *Physiological Reviews* 87: 873–904.

McEwen, B. S. (2016). "In Pursuit of Resilience: Stress, Epigenetics, and Brain Plasticity." *Annals of the New York Academy of Sciences* 1373: 56–64.

McEwen, B. S., and E. Stellar. (1993). "Stress and the Individual: Mechanisms Leading to Disease." *Archives of Internal Medicine* 153: 2093–3101.

McGaugh, J. L. (2000). "Memory: A Century of Consolidation." *Science* 287: 248–251.

McHugh, P. R., and P. R. Slavney. ([1983] 1986). *The Perspectives of Psychiatry*. Baltimore: Johns Hopkins University Press.

Meaney, M. J. (2001). "Maternal Care, Gene Expression, and the Transmission of Individual Differences in Stress Reactivity Across Generations." *Annual Review of Neuroscience* 24: 1161–1192.

Mellars, P. (2004). "Neanderthals and the Modern Colonization of Europe." *Nature* 432: 461–465.

Mellars, P. (2006). "Why Did Modern Human Populations Disperse from Africa ca. 60,000 Years Ago?" *Proceedings of the National Academy of Sciences of the United States of America* 103: 9381–9386.

Meltzoff, A. N., and M. K. Moore. (1977). "Imitation of Facial and Manual Gestures by Human Neonates." *Science* 198: 75–78.

Merton, P. A., and H. B. Morton. (1980). "Stimulation of the Cerebral Cortex in the Intact Human Subject." *Nature* 285: 227–230.

Mess, A., and A. M. Carter. (2007). "Evolution of the Placenta During the Early Radiation of Placental Mammals." *Comparative Biochemistry and Physiology, Part A: Molecular and Integrative Physiology* 148: 769–779.

Meyer-Lindenberg, A. (2008). "Impact of Prosocial Neuropeptides on Human Brain Function." *Progress in Brain Research* 170: 463–470.

Meynert, T. (1872). "Vom Gehirn der Saugetiere." In *Handbuch der lehre von den geweben des menschen und tiere*, ed. S. Stricker, 694–808. Leipzig: Engelmann.

Miller, G. A. (2010). "Mistreating Psychology in the Decades of the Brain." *Perspectives on Psychological Science* 5: 716–743.

Miller, G., E. Galanter, and K. Pribram. (1960). *Plans and the Structure of Behavior*. New York: Henry Holt.

Miller, N. E. (1957). "Experiments of Motivation: Studies Combining Psychological, Physiological, and Pharmacological Techniques." *Science* 126: 1271–1278.

Miller, N. E. (1965). "Chemical Coding of Behavior in the Brain." *Science* 148: 328–338.

Miller, S. M., D. Marcotulli, A. Shen, and L. S. Zweifel. (2019). "Divergent Medial Amygdala Projections Regulate Approach-Avoidance Conflict Behavior." *Nature Neuroscience* 22: 565–575.

Milner, A. D., and M. A. Goodale ([1995] 2000). *The Visual Brain in Action.* Oxford: Oxford University Press.

Ming, G. L., and H. Song. (2005). "Adult Neurogenesis in the Mammalian Central Nervous System."*Annual Review of Neuroscience* 28: 223–250.

Mithen, S. (1996). *The Prehistory of the Mind: The Cognitive Origins of Art and Science.* London: Thames and Hudson.

Mithen, S. (2006). *The Singing Neanderthal: The Origins of Music, Language, Mind and Body.* Cambridge, MA: Harvard University Press.

Miranda, R. A., W. D. Casebeer, A. M. Hein, J. W. Judy, E. P. Krotkov, T. L. Laabs, . . . G. S. Ling. (2014). "DARPA-Funded Efforts in the Development of Novel Brain–Computer Interface Technologies." *Journal of Neuroscience Methods* 244: 52–67.

Mithoefer, M. C., C. S. Grob, and T. D. Brewerton. (2016). "Novel Psychopharmacological Therapies for Psychiatric Disorders: Psilocybin and MDMA." *Lancet Psychiatry* 3: 619–627.

Mitra, R., S. Jadhav, B. S. McEwen, A. Vyas, and S. Chattarji. (2005). "Stress Duration Modulates the Spatiotemporal Patterns of Spine Formation in ihe Basolateral Amygdala." *Proceedings of the National Academy of Sciences of the United States of America* 102: 9371–9376.

Mogenson, G. J. (1987). "Limbic-Motor Integration. In *Progress in Psychobiology and Physiological Psychology*, ed. A. N. Epstein and J. Sprague. New York: Academic Press.

Moll, H., and M. Tomasello. (2007). "Cooperation and Human Cognition: The Vygotskian Intelligence Hypothesis." *Philosophical Transactions of the Royal Society of London, Series B: Biological Sciences* 362: 639–648.

Moll, J., R. de Oliveira-Souza, F. T. Moll, F. A. Ignácio, I. E. Bramati, E. M. Caparelli-Dáquer, and P. J. Eslinger. (2005). "The Moral Affiliations of Disgust: A Functional MRI Study." *Cognitive and Behavioral Neurology* 18: 68–78.

Moll, J., F. Krueger, R. Zahn, M. Pardini, R. de Oliveira-Souza, and J. Grafman. (2006). "Human Fronto-Mesolimbic Networks Guide Decisions About Charitable Donation." *Proceedings of the National Academy of Sciences of the United States of America* 103: 15623–15628.

Moore-Ede, M. C., F. M. Sulzman, and C. A. Fuller. (1992). *The Clocks That Time Us: Physiology of the Circadian Timing System.* Cambridge, MA: Harvard University Press.

Moran, T. H. (2006). "Gut Peptide Signaling in the Controls of Food Intake." *Obesity* 14: 2005–2008.

Moreno, J. D. (1995). *Deciding Together: Bioethics and Moral Consensus.* Oxford: Oxford University Press.

Moreno, J. D. (2011). *The Bodily Politic: The Battle Over Science in America.* New York: Bellevue Literary Press.

Moreno, J. D. (2012). *Mind Wars: Brain Science and the Military in the 21st Century.* New York: Bellevue Literary Press.

Morgan, M. A., J. Schulkin, and J. E. LeDoux. (2003). "Ventral Medial Prefrontal Cortex and Emotional Perseveration: The Memory for Prior Extinction Training." *Behavioral Brain Research* 146: 121–30.

Morgan, T. (1925). *Evolution and Genetics.* Princeton, NJ: Princeton University Press.

Morris, J. S., A. Ohman, and R. J. Dolan. (1999). "A Subcortical Pathway to the Right Amygdala Mediating 'Unseen' Fear." *Proceedings of the National Academy of Sciences of the United States of America* 96: 1680–1685.

Morrison, A. (1993). "Symposium: Dream Research Methodology— Mechanisms Underlying Oneiric Behavior Released in REM Sleep by Pontine Lesions in Cats." *Journal of Sleep Research* 2: 4–7.

Morrow, E. M., S. Y. Yoo, S. W. Flavell, T. K. Kim, Y. Lin, R. S. Hill, . . . C. A. Walsh. (2008). "Identifying Autism Loci and Genes by Tracing Recent Shared Ancestry." *Science* 321: 218–223.

Moruzzi, G., and H. Magoun. (1949). "Brain Stem Reticular Formation and Activation of the EEG." *Electroencephalography and Clinical Neurophysiology* 1: 455–473.

Mota, M. T., and M. B. C. Sousa. (2000). "Prolactin Levels of Fathers and Helpers Related to Alloparental Care in Common Marmosets, *Callithrix jacchus.*" *Folia Primatology* 71: 22–26.

Mountcastle, V. B. (1998). *Perceptual Systems: The Cerebral Cortex.* Cambridge, MA: Harvard University Press.

Murray, E. A., and S. P. Wise. (2004). "What, If Anything, Is the Medial Temporal Lobe, and How Can the Amygdala Be Part of It If There Is No Such Thing?" *Neurobiology of Learning and Memory* 82: 178–198.

Mustard, J. A., K. T. Beggs, and A. R. Mercer. (2005). "Molecular Biology of the Invertebrate Dopamine Receptors." *Archives of Insect Biochemistry and Physiology* 59: 103–117.

Nagel, T. (1979). "What Is It Like to Be a Bat?" In *Mortal Questions*, ed. E. Nagel, 165–180. Cambridge: Cambridge University Press.

Naselaris, T., K. N. Kay, S. Nishimoto, and J. L. Gallant. (2011). "Encoding and Decoding in fMRI." *Neuroimage* 56: 400–410.

National Research Council. 2008. *Emerging Cognitive Neuroscience and Related Technologies*, Washington, D.C.: National Academies Press.

Nature. (2003). "Silence of the Neuroengineers." Editorial. *Nature* 423: 787. doi:https://doi.org/10.1038/423787b.

Nauta, W. J. H. (1972). "The Central Visceromotor System: A General Survey." In *Limbic System Mechanisms and Autonomic Function*, ed. C. H. Hockman. Springfield, IL: Charles C. Thomas.

Nauta, W. J. H., and V. B. Domesick. (1982). "Neural Associations of the Limbic System." In *The Neural Basis of Behavior*, ed. A. L. Beckman. New York: Spectrum.

Nave, G., A. Nadler, D. Dubois, D. Zava, C. Camerer, and H. Plassmann. 2018. "Single-Dose Testosterone Administration Increases Men's Preference for Status Goods." *Nature Communications* 9. https://www.nature.com/articles/s41467-018-04923-0.

Neurath, O. (1944). *Foundations of the Social Sciences*. Chicago: University of Chicago Press.

Nitabach, M., J. Blau, and T. Holmes. (2002). "Electrical Silencing of *Drosophila* Pacemaker Neurons Stops the Free-Running Circadian Clock." *Cell* 109: 485–495.

Norgren, R. (1995). "Gustatory System." In *The Rat Nervous System*, 2nd ed., ed. G. Paxnios, 751–771. San Diego, CA: Academic Press.

Nottebohm, F. (1994). "The Song Circuits of the Avian Brain as a Model System in Which to Study Vocal Learning, Communication and Manipulation." *Discussions in Neurosciences* 10: 72–81.

NeuroFocus. 2011. "NeuroFocus Announces World's First Wireless Full-Brain EEG Measurement Headset: Mynd™." *PR Newswire*, March 21, 2011. https://www.prnewswire.com/news-releases/neurofocus-announces-worlds-first-wireless-full-brain-eeg-measurement-headset--mynd-118355014.html.

Nishimoto, S., A. T. Vu, T. Naselaris, Y. Benjamini, B. Yu, and J. L. Gallant. 2011. "Reconstructing Visual Experiences from Brain Activity Evoked by Natural Movies." *Current Biology* 21 (19): 1641–1646.

Nugent, A. C., E. D. Ballard, T. D. Gould, L. T. Park, R. Moaddel, N. E. Brutsche, and C. A. Zarate Jr. (2018). "Ketamine Has Distinct Electrophysiological and Behavioral Effects in Depressed and Healthy Subjects." *Molecular Psychiatry* 27: 1038–1048.

O'Doherty, J., P. Dayan, J. Schultz, R. Deichmann, K. Friston, and R. J. Dolan. (2004). "Dissociable Roles of Ventral and Dorsal Striatum in Instrumental Conditioning." *Science* 304: 452–455.

O'Keefe, J., and L. Nadel. (1979). "The Hippocampus as a Cognitive Map." *Behavioral and Brain Sciences* 2: 487–533.

Okum, M. (2014). "Deep-Brain Stimulation—Entering the Era of Human Neural-Network Stimulation." *New England Journal of Medicine* 371: 1369–1373.

Olazabal, D. E., and L. J. Young. (2006). "Species and Individual Differences in Juvenile Female Alloparental Care Are Associated with Oxytocin Receptor Density in the Striatum and the Lateral Septum." *Hormones and Behavior* 49: 681–687.

Olds, J. (1958). "Self-Stimulation of the Brain." *Science* 127: 315–324.

Olsson, A., and E. A. Phelps. (2007). "Social Learning of Fear." *Nature Neuroscience* 10: 1095–1102.

Olton, D., J. Becker, and G. E. Handelmann. (1979). "Hippocampus, Space, and Memory." *Behavioral and Brain Sciences* 2: 313–365.

Osmundsen, J. A. (1965). "'Matador' with a Radio Stops Wired Bull: Modified Behavior in Animals Subject of Brain Study." *New York Times*, May 17, 1965.

Paabo, S. (2015). *Neanderthal Man: In Search of Lost Genes*. New York: Basic Books.

Paabo, S., H. Poinar, D. Serre, V. Jaenicke-Despres, J. Hebler, N. Rohland, M. Kuch, J. Krause, L. Vigilant, and M. Hofreiter. (2004). Genetic Analyses from Ancient DNA." *Annual Review of Genetics* 38: 645–679.

Pais-Vieira, M., G. Chiuffa, M. Lebedev, A. Yadav, and M. A. Nicolelis. (2015). "Building an Organic Computing Device with Multiple Interconnected Brains." *Scientific Reports* 5: 11869.

Palombit, R. A., R. M. Seyfarth, and D. L. Cheney. (1997). "The Adaptive Value of "Friendships" to Female Baboons: Experimental and Observational Evidence." *Animal Behavior* 54: 599–614.

Pandarinath, C., P. Nuyujukian, C. H. Blabe, B. L. Sorice, J. Saab, F. R. Willett, L. R. Hochberg, K. V. Shenoy, and J. M. Henderson. (2017).

"High Performance Communication by People with Paralysis Using an Intracortical Brain-Computer Interface." *ELife*. DOI:10.7554/eLife .18554.

Panksepp, J. (1998). *Affective Neuroscience: The Foundations of Human and Animal Emotions*. New York: Oxford University Press.

Papez, J. W. (1937). "A Proposed Mechanism of Emotion." *Archives of Neurological Psychiatry* 38: 725–743.

Park, H. G., and K. Friston. (2013). "Structural and Functional Brain Networks: From Connections to Cognition." *Science* 342: 579–584.

Partridge, J. G., P. A. Forcelli, R. Luo, J. M. Cashdan, J. Schulkin, R. J. Valentino, and S. Vicini. 2016. "Stress Increases GABAergic Neurotransmission in CRF Neurons of the Central Nucleus and the Bed Nucleus of the Stria Terminalis." *Neuropharmacology* 107: 239–250.

Passingham, R. E. ([1993] 1997). *The Frontal Lobes and Voluntary Action*. Oxford: Oxford University Press.

Pavlov, I. P. ([1897] 1902). *The Work of the Digestive Glands*. London: Charles Griffin.

Pavlov, I. P. ([1927] 1960). *Lectures on Conditioned Reflexes*. New York: International Publishing.

Peciña, S., K. S. Smith, and K. C. Berridge. (2006). "Hedonic Hot Spots in the Brain." *Neuroscientist* 12: 500–511.

Peirce, C. S. (1877). "The Fixation of Belief." *Popular Science Monthly* 12: 1–15.

Peirce, C. S. ([1899] 1992). *Reasoning and the Logic of Things*. Ed. K. L. Ketner and H. Putnam. Cambridge, MA: Harvard University Press.

Penfield, W., and T. Rasmussen. (1950). *The Cerebral Cortex of Man*. New York: Macmillan.

Perret, D. I., and N. J. Emery. (1994). "Understanding the Intentions of Others from Visual Signals: Neurophysiological Evidence." *Cahiers de pscyhologie cognitive* 13: 683–694.

Perrett, D., E. T. Rolls, and W. Caan. (1982). "Visual Neurones Responsive to Faces in the Monkey Temporal Cortex." *Experimental Brain Research* 47: 329–342.

Perry, C. J., L. Baciadonna, and L. Chittka. (2016). "Unexpected Rewards Induce Dopamine-Dependent Position Emotion-like State Changes in Bumblebees." *Science* 353: 1529–1531.

Peters, E., K. McCaul, M. Stefanek, and W. Nelson. (2006). "A Heuristics Approach to Understanding Cancer Risk Perception: Contributions from

Judgment and Decision-Making Research." *Annals of Behavioral Medicine* 31: 45–52.

Petrovic, P., R. Kalisch, T. Singer, and R. J. Dolan. (2008). "Oxytocin Attenuates Affective Evaluations of Conditioned Faces and Amygdala Activity." *Journal of Neuroscience* 28: 6607–6615.

Pfaff, D. (1980). *Estrogens and Brain Function: Neural Analysis of a Hormone-Controlled Mammalian Reproductive Behavior.* New York: Springer-Verlag.

Pfaff, D., L. Westberg, and L. Kow. (2005). "Generalized Arousal of Mammalian Central Nervous System." *Journal of Comparative Neurology* 493: 86–91.

Pfaffmann, C., R. Norgren, and H. J. Grill. (1977). "Sensory Affect and Motivation." *Annals of the New York Academy of Sciences* 290: 18–34.

Phelps, E., K. O'Connor, W. Cunningham, E. Funayama, J. Gatenby, J. Gore, and M. Banaji. (2000). "Performance on Indirect Measures of Race Evaluation Predicts Amygdala Activation." *Journal of Comparative Neurology* 12: 729–738.

Piccolino, M. (1997). "Luigi Galvani and Animal Electricity: Two Centuries After the Foundation of Electrophysiology." *Trends in Neuroscience* 20: 443–448.

Pinnock, S. B., A. M. Blake, N. J. Platt, and J. Herbert. (2010). "The Role of BDNF, pCREB, and Wnt3a in the Latent Period Preceding Activation of Progenitor Cell Mitosis in the Dentate Gyrus by Fluoxetine." *PLoS One* 5: e13652.

Pinker, S. E. (2018). *Enlightenment Now: The Case for Reason, Science, Humanism, and Progress.* New York. Random House.

Plassmann, H., J. O'Doherty, B. Shiv, and A. Rangel. 2008. "Marketing Actions Can Modulate Neural Representations of Experienced Pleasantness." *Proceedings of the National Academy of Sciences* 105 (3): 1050–1054.

Poldrack, R. A. (2018). *The New Mind Readers: What Neuroimaging Can and Cannot Reveal About Our Thoughts.* Princeton, NJ: Princeton University Press.

Pomrenze, M. B., J. Tovar-Diaz, A. Blasio, R. Maiya, S. M. Giovanetti, K. Lei, H. Morikawa, F. Woodward Hopf, and R. O. Messing. (2019). "A Corticotropin Releasing Factor Network in the Extended Amygdala for Anxiety." *Journal of Neuroscience* 39: 3030–3043.

Porto, P. R., L. Oliveira, J. Mari, E. Volchan, I. Figueira, and P. Ventura. (2009). "Does Cognitive Behavioral Therapy Change the Brain? A Systematic Review of Neuroimaging in Anxiety Disorders." *Journal of Neuropsychiatry and Clinical Neuroscience* 21: 114–125.

Powley, T. L. (2000). "Vagal Input to the Enteric Nervous System." *Gut* 47 (supplement 4): iv30–iv32.

Premack, D., and A. J. Premack. (1995). "Origins of Human Social Competence." In *The Cognitive Neurosciences*, ed. M. S. Gazzaniga, 205–218. Cambridge, MA: MIT Press.

Preuss, T. (1993). "The Role of Neurosciences in Primate Evolutionary Biology: Historical Commentary and Prospectus." In *Primates and Their Relatives in Phylogenetic Perspective*, ed. R. D. E. MacPhee. New York: Springer.

Pulvermuller, F., O. Hauk, V. V. Nikulin, and R. J. Ilmoniermi. (2005). "Functional Links Between Motor and Language Systems." *European Journal of Neuroscience* 21: 793–797.

Quine, W. V. O. ([1953] 1961). *From a Logical Point of View: Nine Logico-Philosophical Essays*. New York: Harper Torchbooks.

Raichle, M. E. (2015). "The Brain's Default Mode Network." *Annual Review of Neuroscience* 38: 433–447.

Rakic, P. (2002). "Evolving Concepts of Cortical Radial and Areal Specification." *Progress in Brain Research* 136: 265–280.

Rakic, P. (2009). "Evolution of the Neocortex: Perspective from Developmental Biology." *Nature Reviews: Neuroscience* 10: 724–735.

Ramachandran, V. (1999). *Phantoms in the Brain: Probing the Mysteries of the Human Mind*. New York: HarperCollins.

Rao, R. P., A. Stocco, M. Bryan, D. Sarma, T. M. Youngquist, J. Wu, and C. S. Prat. (2014). "A Direct Brain-to-Brain Interface in Humans." *PLoS One* 9.

Rasgon, N. L., and B. S. McEwen. (2016). "Insulin Resistance—A Missing Link No More." *Molecular Psychiatry* 21: 1648–1652.

Rauschecker, J. P. (2012). "Ventral and Dorsal Streams in the Evolution of Speech and Language." *Frontiers in Evolutionary Neuroscience* 4: 1–4.

Rauschecker, J. P., and S. K. Scott. (2009). "Maps and Streams in the Auditory Cortex: Nonhuman Primates Illuminates Human Speech Processing." *Nature Neuroscience* 12: 718–724.

Reader, S. M., and K. N. Laland. (2002). "Social Intelligence, Innovation, and Enhanced Brain Size in Primates." *Proceedings of the National Academy of Sciences of the United States of America* 99: 4436–4441.

Reep, R. L., B. L. Finlay, and R. B. Darlington. (2007). "The Limbic System in Mammalian Brain Evolution." *Brain, Behavior, Evolution* 70: 57070.

Rescorla, R. A., and A. R. Wanger. (1972). "A Theory of Pavlovian Conditioning: Variations in the Effectiveness of Reinforcement and Nonreinforcment." In *Classical Conditioning: Current Research and Theory*, ed. W. J. Baker and W. Prokasy. New York: Appleton-Century-Crofts.

Ressler, K. J., and H. S. Mayberg. (2007). "Targeting Abnormal Neural Circuits in Mood and Anxiety Disorders: From the Laboratory to the Clinic." *Nature Neuroscience* 10: 1116–1124.

Reynolds, M. 2017. "DeepMind's AI Beats World's Best Go player in Latest Face-Off." *New Scientist*, May 23, 2017. https://www.newscientist.com/article/2132086-deepminds-ai-beats-worlds-best-go-player-in-latest-face-off/.

Richerson, P. J., and R. Boyd. (2005). *Not by Genes Alone: How Culture Transformed Human Evolution*. Chicago: University of Chicago Press.

Richter, C. P. (1943). *Total Self-Regulatory Functions in Animals and Man*. New York: Harvey Lecture Series.

Richter, C. P. (1952). "The Domestic of the Norway Rat and the Implications for the Study of Genetics in Man." *American Journal of Human Genetics* 4: 273–285.

Richter, C. P. ([1965] 1979). *Biological Clocks in Medicine and Psychiatry*. Springfield, IL: Charles C. Thomas.

Rilling, J., A. DeMarco, P. Hackett, R. Thompson, B. Ditzen, R. Patel, and G. Pagnoni. (2012). "Effects of Intranasal Oxytocin and Vasopressin on Cooperative Behavior and Associated Brain Activity in Men." *Psychoneuroendocrinology* 37: 447–461.

Rizzolatti, G., and G. Luppino. (2001). "The Cortical Motor System." *Neuron* 31: 889–901.

Robertson, C. E., E. M. Ratai, and N. Kanwisher. (2016). "Reduced GAGAergic Action in the Autistic Brain." *Current Biology* 26: 1–6.

Rohde, J. (2013). *Armed with Expertise: The Militarization of American Social Research During the Cold War*. New York: Cornell University Press.

Rolls, E. T. (1996). "The Orbitofrontal Cortex." *Philosophical transactions of the Royal Society of London* 351: 1433–1444.

Rolls, E. T. (2000). "Functions of the Primate Temporal Lobe Cortical Visual Areas in Invariant Visual Object and Face Recognition." *Neuron* 27: 205–218.

Rolls, E. T., and G. Deco. (2002). *Computational Neuroscience of Vision.* Oxford: Oxford University Press.

Rose, N. (2014). "The Human Brain Project: Social and Ethical Challenges." *Neuron* 82: 1212–1215.

Rosen, J., and J. Schulkin. (1998). "From Normal Fear to Pathological Anxiety." *Psychological Review* 105: 325–350.

Rosenzweig, M. (1984). "Experience, Memory, and the Brain." *American Psychology* 39: 365–376.

Rousseau, J.-J. ([1782] 1992). *The Reveries of the Solitary Walker.* New York: Hackett.

Rozin, P. (1976). "The Evolution of Intelligence and Access to the Cognitive Unconscious." In *Progress in Psychobiology and Physiological Psychology*, ed. J. Sprague and A. N. Epstein. New York: Academic Press.

Rozin, P. (1999). "The Process of Moralization." *Psychological Science* 10: 218–221.

Rusconi, E., and T. Mitchener-Nissen. (2014). "The Role of Expectations, Hype, and Ethics in Neuroimaging and Neuromodulation Futures." *Frontiers in Systems Neuroscience* 8: 1–7.

Sanford, C. A., M. E. Soden, M. A. Baird, S. M. Miller, J. Schulkin, R. D. Palmiter, M. Clark, and L. S. Zweifel. (2017). "A Central Amygdala CRF Circut Facilitates Learning About Weak Threats." *Neuron* 93 (1): 164–178.

Satel, S., and S. Lilienfeld. (2013). *Brainwashed: The Seductive Appeal of Mindless Neuroscience.* New York: Basic Books.

Schacter, D. L. (1996). *Searching for Memory: The Brain, the Mind, and the Past.* New York: Basic Books.

Schacter, D. L. (2001). *The Seven Sins of Memory: How the Mind Forgets and Remembers.* New York: Houghton.

Schaffer, A. (2006). "It May Come as a Shock." *New York Times*, November 7, 2006. https://www.nytimes.com/2006/11/07/health/07migr.html.

Schubotz, R. I. (2007). "Prediction of External Events with Our Motor System: Towards a New Framework." *Trends in Cognitive Science* 11: 212–218.

Schulkin, J. (2000). *Roots of Social Sensibility and Neural Function.* Cambridge, MA: MIT Press.

Schulkin, J. (2003). *Rethinking Homeostasis: Allostatic Regulation in Physiology and Pathophysiology.* Cambridge, MA: MIT Press.

Schulkin, J. (2017). *The CRF Signal: Uncovering an Information Molecule.* Oxford: Oxford University Press.

Schultz, W. (2007). "Multiple Dopamine Functions at Different Time Courses." *Annual Review of Neuroscience* 30: 59–88.

Schultz, W. (2010). "Dopamine Signals for Reward Value and Risk." *Behavioral and Brain Functions* 6: 1–9.

Schultz, W., K. Preuschoff, C. Camerer, M. Hsu, C. D. Fiorillo, P. N. Tobler, and P. Bossaerts. (2008). "Explicit Neural Signals Reflecting Reward Uncertainty." *Philosophical Transactions of the Royal Society of London, Series B: Biological Sciences* 363: 3801–3811.

Schwartz, C., C. Wright, L. Shin, J. Kagan, P. Whalen, K. McMullin, and S. Rauch. (2003). "Differential Amygdalar Response to Novel Versus Newly Familiar Neutral Faces: A Functional MRI Probe Developed for Studying Inhibited Temperament." *Biological Psychiatry* 53: 854–862.

Searle, J. R. (1992). *The Rediscovery of the Mind.* Cambridge, MA: MIT Press.

Sejnowski, T. (2007). "The Hippocampus Review." *Science* 6: 44–45.

Sellars, W. (1962). *Science, Perception, and Reality.* New York: Routledge.

Seung, H. (2012). *Connectome: How the Brain's Wiring Makes Us Who We Are.* New York: Houghton Mifflin Harcourt.

Shachar-Dadon, A., J. Schulkin, and M. Leshem. (2009). "Adversity Before Conception Will Affect Adult Progeny in Rats." *Developmental Psychology* 45: 9–16.

Shanahan, John. 2019. "Statement by Lieutenant General John N. T. Shanahan on Artificial Intelligence Initiatives." Senate Armed Services Committee, Subcommittee on Emerging Threats and Capabilitiess. March 12, 2019. https://www.armed-services.senate.gov/imo/media/doc/Shanahan_03-12-19.pdf.

Shapin, S. (1996). *The Scientific Revolution.* Chicago: University of Chicago Press.

Shapin, S., and S. Schaffer. (2011). *Leviathan and the Air-Pump: Hobbes, Boyle, and the Experimental Life.* Princeton, NJ: Princeton University Press.

Shepherd, G. (2010). *Creating Modern Neuroscience: The Revolutionary 1950s.* Oxford: Oxford University Press.

Sherbondy, A. J., R. F. Dougherty, R. Ananthanarayanan, D. S. Modha, and B. A. Wandell. (2009). "Think Global, Act Local: Projectome Estimation with BlueMatter." *Medical Image Computing and Computer-Assisted Intervention* 12: 681–688.

Sherrington, C. S. ([1906] 1948). *The Integrative Action of the Nervous System.* New Haven, CT: Yale University Press.

Sherwood, N. M., and D. B. Parker. (1990). "Neuropeptide Families: An Evolutionary Perspective." *Journal of Experimental Zoology* (supplement), 4: 63–71.

Shettleworth, S. J. (1998). *Cognition, Evolution, and Behavior.* Oxford: Oxford University Press.

Shors, T., C. Micseages, A. Beylin, M. Zhao, T. Rydel, and E. Gould. (2001). Neurogenesis in the Adult Is Involved in the Formation of Trace Memories." *Nature* 410: 372–375.

Shultz, S., and R. Dunbar. (2010). "Encephalization Is Not a Universal Macroevolutionary Phenomenon in Mammals But Is Associated with Sociality." *Proceedings of the National Academy of Sciences of the United States of America* 107: 21582–21586.

Silk, J. B. (2007). "The Adaptive Value of Sociality in Mammalian Groups." *Philosophical Transactions of the Royal Society of London, Series B: Biological Sciences* 362: 539–559.

Silver, D., J. Schrittwieser, K. Simonyan, I. Antonoglou, A. Huang, A. Guez, T. Hubert, L. Baker, M. Lai, A. Bolton, Y. Chen, T. Lillicap, F. Hui, L. Sifre, G. van den Driessche, and D. Hassabis. (2017). "Mastering the Game of Go Without Human Knowledge." *Nature* 550: 354–359.

Simon, H. A. (1962). "The Architecture of Complexity." *Proceedings of the American Philosophical Society* 106: 470–473.

Simon, H. A. (1982). *Models of Bounded Rationality.* Cambridge, MA: MIT Press.

Simpson, G. G. (1949). *The Meaning of Evolution: A Study of the History of Life and of Its Significance for Man.* New Haven, CT: Yale University Press.

Sinnott-Armstrong, W. (2008). *Moral Psychology.* Vols. 1–3. Cambridge, MA: MIT Press.

Skinner, B. F. (1976). *About Behaviorism.* New York: Vintage.

Smith, A. ([1759] 1882). *The Theory of Moral Sentiments.* Indianapolis: Liberty Classics.

Spector, A. (2000). "Linking Gustatory Neurobiology to Behavior in Vertebrates." *Neuroscience and Biobehavioral Reviews* 24: 391–416.

Sperry, R. (1965). "Embryogenesis of Behavioral Nerve Nets." In *Organogenesis*, ed. R. I. Deehan and H. Ursprung. New York: Holt, Rinehart and Winston.

Sperry, R. (1961). "The Cerebral Organization of Behavior." *Science* 133: 1749–1757.

Spor, A., O. Koren, and R. Ley. (2011). "Unravelling the Effects of the Environment and Host Genotype on the Gut Microbiome." *Nature Reviews Microbiology* 9: 279–290.

Sporns, O. (2011). *Networks of the Brain*. Cambridge, MA: MIT Press.

Squire, L. R. (2004). "Memory Systems of the Brain: A Brief History and Current Perspective." *Neurobiology of Learning and Memory* 82: 171–177.

Stellar, E. (1954). "The Physiology of Motivation." *Psychological Review* 61: 5–22.

Sterling, P. (2004). "Principles of Allostasis: Optimal Design, Predictive Regulation, Psychopathology, and Rational Therapeutics." In *Allostasis, Homeostasis, and the Costs of Physiological Adaptation*, ed. J. Schulkin. Cambridge: Cambridge University Press.

Sterling, P., and Laughlin, S. (2015). *Principles of neural design*. Cambridge, MA: MIT Press.

Stocco, A., C. S. Prat, D. M. Losey, J. A. Cronin, J. Wu, J. A. Abernethy, and R. P. N. Rao. (2015). "Playing 20 Questions with the Mind: Collaborative Problem Solving by Humans Using a Brain-to-Brain Interface." *PLoS One* (September 23, 2015).

Strand, F. L. (1999). *Neuropeptides: Regulators of Physiological Processes*. Cambridge, MA: MIT Press.

Stumpf, W. E., M. Sar, S. A. Clark, and H. F. DeLuca. (1982). "Brain Target Sites for 1,25-Dihydroxyvitamin D3." *Science* 215: 1403–1405.

Swanson, L. W. (1999). "The Neuroanatomy Revolution of the 1970s and the Hypothalamus." *Brain Research Bulletin* 50: 397.

Swanson, L. W. (2000). "Cerebral Hemisphere Regulation of Motivated Behavior." *Brain Research* 886: 113–164.

Swanson, L. W. ([2011] 2015). *Brain Architecture: Understanding the Basic Plan*. 2nd ed. New York: Oxford University Press.

Swanson, L. W., and J. F. Lichtman. (2016). "From Cajal to Connectome and Beyond." *Annual Review of Neuroscience* 39: 197–216.

Tager-Flusberg, H., ed. (1999). *Neurodevelopmental Disorders.* Cambridge, MA: MIT Press.

Tallis, R. (2014). *Aping Mankind: Neuromania, Darwinitis, and the Misrepresentation of Humanity.* London: Routledge.

Tang-Schomer, M. D., J. D. White, L. W. Tien, L. I. Schmitt, T. M. Valentin, D. J. Graziano, A. M. Hopkins, F. G. Omenetto, P. G. Haydon, and D. L. Kaplan. (2014). "Bioengineered Functional Brain-like Cortical Tissue." *Proceedings of the National Academy of Sciences of the United States of America* 111 (38): 13811–13816.

Tegmark, M. (2017). *Life 3.0: Being Human in the Age of Artificial Intelligence.* New York: Knopf.

Tenenbaum, J., C. Kemp, T. Griffiths, and N. Goodman. (2011). "How to Grow a Mind: Statistics, Structure, and Abstraction." *Science* 331: 1279–1285.

Thorpe, S. K. S., R. L. Holder, and R. H. Crompton. (2007). "Origin of Human Bipedalism as an Adaptation for Locomotion on Flexible Branches." *Science* 316: 1328–1331.

Tinbergen, N. ([1951] 1969). *The Study of Instinct.* Oxford: Oxford University Press.

Todes, D. P. (2014). *Ivan Pavlov: A Russian Life in Science.* New York: Oxford University Press.

Tomasello, M. (1999). *The Cultural Origins of Human Cognition.* Cambridge, MA: Harvard University Press.

Tomasello, M. (2009). *Why We Cooperate.* Cambridge, MA: MIT Press.

Tononi, G., M. Boly, M. Massimini, and C. Koch. (2016). "Integrated Information Theory: From Consciousness to Its Physical Substrate." *Nature Reviews: Neuroscience* 17: 450–461.

Toulmin, S. ([1950] 1970). *Reason in Ethics.* Cambridge: Cambridge University Press.

Toulmin, S. (1977). *Human Understanding.* Princeton, NJ: Princeton University Press.

Toyama, K., S. Matsumoto, M. Kurasawa, H. Setoguchi, T. Noma, K. Takenaka, A. Soeda, M. Shimodozono, and K. Kawahira. 2014. "Novel Neuromuscular Electrical Stimulation System for Treatment of Dysphagia After Brain Injury." *Neurologia medico-chirurgica* 54 (7): 521–528.

Tyler, W. J. (2011). "Noninvasive Neuromodulation with Ultrasound." *Neuroscientist* 17: 25-36.

Tulving, E., and F. I. M. Craik. (2000). *The Oxford Handbook of Memory*. Oxford: Oxford University Press.

Turing, A. (2004). *The Essential Turing: Seminal Writings in Computing, Logic, Philosophy, Artificial Intelligence, and Artificial Life plus The Secrets of Enigma*. Ed. B. J. Copeland. Oxford: Clarendon.

Turk-Browne, N. B. (2013). "Functional Interactions as Big Data in the Human Brain." *Science* 342: 580–584.

Ullman, M. T. (2001). "A Neurocognitive Perspective on Language: The Declarative/Procedural Model." *Nature Reviews: Neuroscience* 9: 266–286.

Ullman, M. T. (2004). "Is Broca's Area Part of a Basal Ganglia Thalamocortical Circuit?" *Cognition* 92: 231–270.

Ungerleider, L. G., and M. Mishkin. (1982). "Two Cortical Visual Systems." In *Analysis of Visual Behavior*, ed. D. Ingle, M. Goodale, and R. Mansfield. Cambridge, MA: MIT Press.

Valenstein, E. (1973). *Brain Control*. New York: Wiley.

Valenstein, E. (2006). *The War of the Soups and the Sparks: The Discovery of Neurotransmitters and the Dispute Over How Nerves Communicate*. New York: Columbia University Press.

Van Essen, D. C. (2005). "Corticocortical and Thalamocortical Information Flow in the Primate Visual System." *Progress in Brain Research* 149: 173–185.

Van Essen, D. C., C. H. Anderson, and D. J. Felleman. (1992). "Information Processing in the Primate Visual System: An Integrated Systems Perspective." *Science* 255: 419–422.

Vecchiato, G., L. Astolfi, F. De Vico Fallani, J. Toppi, F. Aloise, F. Bez, D. Wei, W. Kong, J. Dai, F. Cincotti, D. Mattia, and F. Babiloni. (2011). "On the Use of EEG or MEG Brain Imaging Tools in Neuromarketing Research." *Computational Intelligence and Neuroscience* 2011. DOI: 10 .1155/2011/643489.

Veenema, A., and I. Neumann. (2008). "Central Vasopressin and Oxytocin Release: Regulation of Complex Social Disorders." *Progress in Brain Research* 170: 261–276.

Verstynen, T., and B. Voytek. (2014). *Do Zombies Dream of Undead Sheep? A Neuroscientific View of the Zombie Brain*. Princeton, NJ: Princeton University Press.

Volkow, N. D., G. F. Koob, and T. McLellan. (2016). "Neurobiologic Advances from the Brain Disease Model of Addiction." *New England Journal of Medicine* 374: 363–371.

Vrselja, Z., S. G. Daniele, J. Silbereis, F. Talpo, Y. M. Morozov, A. M. M. Sousa, B. S. Tanaka, M. Skarica, M. Pletikos, N. Kaur, Z. W. Zhuang, Z. Liu, R. Alkawadri, A. J. Sinusas, S. R. Latham, S. G. Waxman, and N. Sestan. (2019). "Restoration of Brain Circulation and Cellular Functions Hours Post-Mortem." *Nature* 568: 336–343.

Wander, J. D., T. Blakely, K. J. Miller, K. E. Weaver, L. A. Johnson, J. D. Olson, E. E. Fetz, R. P. N. Rao, and J. G. Ojemann. (2013). "Distributed Cortical Adaptation During Learning of a Brain-Computer Interface Task." *Proceedings of the National Academy of Sciences of the United States of America* 110: 10818–10823.

Wander, J. D., and R. P. N. Rao. (2014). "Brain-Computer Interfaces: A Powerful Tool for Scientific Inquiry." *Current Opinion in Neurobiology* 25: 70–75.

Warwick, K., M. Gasson, B. Hutt, I. Goodhew, P. Kyberd, B. Andrews, P. Teddy, and A. Shad. (2003). "The Application of Implant Technology for Cybernetic Systems." *Archives of Neurology* 60 (10): 1369–1373.

Warwick, K., M. Gasson, B. Hutt, I. Goodhew, P. Kyberd, H. Schulzrinne, and X. Wu. (2004). "Thought Communication and Control: A First Step Using Radiotelegraphy." *IEE Proceedings on Communications* 151 (3): 185-189.

Watson, P. (1978). *War on the Mind: The Military Uses and Abuses of Psychology.* New York: Basic Books.

Weaver, I. C., N. Cervoni, F. A. Champagne, A. C. D'Alessio, S. Sharma, J. R. Seckl, S. Dymov, M. Szyf, and M. J. Meaney. (2004). "Epigenetic Programming by Maternal Behavior." *Nature Neuroscience* 7: 847–854.

Weisberg, S., F. Keil, J. Goodstein, E. Rawson, and R. Gray. (2008). "Seductive Allure of Neuroscience Explanations." *Journal of Cognitive Neuroscience* 20: 470–477.

Weiss, P. A. (1939). *Principles of Development.* New York: Henry Holt.

Whitehead, A. N. (1926). *Science and the Modern World.* Cambridge: Cambridge University Press.

Whiten, A., and R. Byrne. (1988). "Tactical Deception in Primates." *Behavioral and Brain Sciences* 11: 233–273.

Wilkins, A. S., R. W. Wrangham, and W. T. Fitch. (2014). The Domestication Syndrome in Mammals: A Unified Explanation Based on Neural Crest Cell Behavior and Genetics." *Genetics* 197: 795–808.

Wimmer, G., and D. Shohamy. (2012). "Preference by Association: How Memory Mechanisms in the Hippocampus Bias Decisions." *Science* 338: 270–273.

Wise, R. A. (2005). "Forebrain Substrates of Reward and Motivation." *Journal of Comparative Neurology* 493: 115–121.

Wittgenstein, L. ([1953] 1968). *Philosophical Investigations*. New York: Macmillan.

Wittmer, C. R., T. Claudepierre, M. Reber, P. Wiedemann, J. A. Garlick, D. Kaplan, and C. Egles. (2011). "Multifunctionalized Electrospun Silk Fibers Promote Axon Regeneration in Central Nervous System." *Advanced Functional Materials* 21: 12–24.

Woods, S. C., R. A. Hutton, and W. Makous. (1970). "Conditioned Insulin Secretion in the Albino Rat." *Proceedings of the Society of Experimental Biology and Medicine* 133: 965–968.

Wynn, T., and F. Coolidge. (2004). "The Expert Neandertal Mind." *Journal of Human Evolution* 46: 1–21.

Yamamoto, K., and P. Vernier. (2011). "The Evolution of Dopamine Systems in Chordates." *Frontiers in Neuroanatomy* 5: 1–21.

Yang, Y., and A. Raine. (2009). "Prefrontal Structural and Functional Brain Imaging Findings in Antisocial, Violent, and Psychopathic Individuals." *Psychiatry Review* 174: 81–88.

Yao, M., J. Schulkin, and R. J. Denver. (2008). "Evolutionary Conserved Glucocorticoid Regulation of CRH." *Endocrinology* 149: 2352–2360.

Young, A. W, ed. (1998). *Face and Mind*. Oxford: Oxford University Press.

Young, R. M. (1970). *Mind, Brain, and Adaptation in the Nineteenth Century*. Oxford: Oxford University Press.

Zak, P. J., R. Kurzban, and W. T. Matzner. (2005). "Oxytocin Is Associated with Human Trustworthiness." *Hormones and Behavior* 48: 522–527.

Zald, D. H., and S. L. Rauch. (2006). *The Orbitofrontal Cortex*. Oxford: Oxford University Press.

Zeki, S., and A. Bartels. (1998). "The Autonomy of the Visual Systems and the Modularity of Conscious Vision." *Philosophical Transactions of the Royal Society of London, Series B: Biological Sciences* 353: 1911–1914.

Zeki, S., and A. Bartels. (1999). Toward a Theory of Visual Consciousness." *Consciousness and Cognition* 8: 225–259.

Zeng, H., E. H. Shen, J. G. Hohmann, S. W. Oh, A. Bernard, J. J. Royall, . . . A. R. Jones. (2012). "Large-Scale Cellular-Resolution Gene Profiling in Human Neocortex Reveals Species Specific Molecular Signatures." *Cell* 149: 483–496.

Zweifel, L. S., J. P. Fadok, E. Argilli, M. G. Garelick, G. L. Jones, T. M. K. Dickerson, J. M. Allen, S. J. Y. Mizumori, A. Bonci, and R. D. Palmiter. (2011). "Activation of Dopamine Neurons Is Critical for Aversive Conditioning and Prevention of Generalized Anxiety." *Nature Neuroscience* 14: 620–626.

INDEX

consciousness, 8, 50–55; AI and, 56–58; brain death and, 54–55; of computer, 56–57; events associated with, 51; hard problem of, 52, 66; James on, 50–52; nature of, 55–56; reproducing, 51; vegetative states, 54

consent, 47

consumerism, testosterone and, 119–120

cooperation, 192

cortex, 15; anterior cingulate, 118; electrical stimulation of, 16–17; evolution of, 71; frontal, 71; motor, 14, 50, 71–72; of mouse, 59; neurons of, 44; occipital, 116; orbitofrontal, 116; prefrontal, 71, 116, 117; premotor, 71–72; visual, 107–108

corticotropin-releasing hormone (CRH), 23–24, 30–31, 79, 131–132

cortisol, 82

CRH. *See* corticotropin-releasing hormone

criminal intention, 113

CT. *See* computer tomography

DARPA. *See* Defense Advanced Research Projects Agency

Darwin, Charles, 29, 192; on disgust, 184; on domestication, 83; evolution and, 68; on morality, 182

Daubert v. Merrell Dow Pharmaceuticals Inc., 115

Davis, Michael, 132

Dawkins, Richard, 35

DBS. *See* deep brain stimulation

deafness, 13

deception, 111–113, 184; brain regions and, 106; imaging, 103–106; neocortical expansion and, 193

deception-detecting systems, 103–105, 111–112

decision-making, 144–145; ethical, 181–182; moral, 187

deep brain stimulation (DBS), 100, 135–136; applications, 154; therapeutic, 172–173

deep learning, 49, 153

Defense Advanced Research Projects Agency (DARPA), 127–128; EEG caps and, 142; Neural Engineering System Design project, 155; neurosecurity funded by, 150; parapsychology, 141–142; Systems-Based Neurotechnology for Emerging Therapies program, 154

deinstitutionalization, 166

Delgado, José, 6–7

dementias, 27

Dennett, Daniel, 57

depression, 28, 114, 165, 171, 173, 179

designer drugs, 133–134

designer receptors exclusively activated by designer drugs (DREADD), 133–135

Dewey, John, 43, 179, 194

diabetes, 27–28